# The History of the
# Space Shuttle

**Whitman**
Publishing, LLC
PUBLISHING SINCE 1934

www.whitman.com

The History of the Space Shuttle

www.whitman.com

ISBN: 0794836585
Printed and assembled in China

To view other products by Whitman Publishing, please visit **Whitman.com**.

# Table of
# **Contents**

Atlantis

# INTRODUCTION

## America's Dependable
## Workhorse

**W**ith its delta-winged orbiter, twin solid rocket boosters and bright orange external fuel tank, NASA's Space Transportation System, better known as the space shuttle, is perhaps the world's most recognizable spacecraft. And since its first launch in 1981, the shuttle has proved to be the dependable, versatile and reusable workhorse it was designed to be.

The shuttle program's five orbiters — *Columbia, Challenger, Discovery, Atlantis* and *Endeavour* — have flown a total of 135 missions over the last 30 years. The shuttle has delivered hundreds of tons of payloads into orbit from its huge cargo bay — military and communications satellites, the Hubble Space Telescope and even an interplanetary spacecraft — while technological breakthroughs such as the Canadarm Remote Manipulator System have helped rescue and repair malfunctioning satellites and build the International Space Station.

The shuttle's design allowed for larger crews, including a number of space firsts. And because the shuttle flew more often than any spacecraft before it — up to eight missions per year in some years — astronauts racked up record numbers of flights, hours in orbit and time spent on spacewalks. As the missions became more routine, crews found more time to enjoy microgravity and the unique experience of being in space.

The shuttle program was not without its share of tragedy, however. Two crews perished when their vehicles failed — *Challenger* on the way to space, and *Columbia* on the way home — and their sacrifice is also part of the shuttle's story. That story came to an end in the summer of 2011, when the final shuttle mission landed safely. But the space shuttle has earned its place in history by immeasurably advancing our knowledge of spaceflight and pushing us ever closer to farther exploration of our universe.

---

(Above) The space shuttle's last mission, delivering a Multi-Purpose Logistics Module full of cargo to the International Space Station, marked the last of 135 missions over 30 years of service. (Opposite page) Whether rocketing toward the heavens on twin pillars of fire or floating in orbit with the Earth as a backdrop, the shuttle has become the most recognizable space vehicle ever built.

**Hear NASA Administrator Charles Bolden**

talk about the legacy
of the space shuttle program

Use your QR Code-Enabled device to see and hear
the sights & sounds of space shuttle history!

# DESIGN AND TESTING

Even before NASA had accomplished President John F. Kennedy's goal of landing a man on the moon, administrators were already discussing the future of the space program post-Apollo. In 1969, the National Aeronautics and Space Council met to consider several options, among them a manned mission to Mars, a follow-on lunar program and a low earth orbital infrastructure program.

Based on the advice of the Council, President Richard M. Nixon made the decision to pursue the low earth orbital infrastructure option, which consisted of construction of a space station and the development of a space shuttle. Nixon announced his decision Jan. 5, 1972, and on April 20, the U.S. House of Representatives passed the space budget — a fact that was relayed to Apollo 16 astronauts John Young and Charlie Duke on the surface of the moon. Young, little knowing he would command the shuttle's first mission, replied "The country needs that shuttle mighty bad."

NASA continued to use disposable Apollo hardware for Skylab, the agency's first Earth-orbiting space station, and the Apollo-Soyuz Test Project. But plans were now in the works for a reusable spacecraft, which NASA hoped would significantly reduce the cost of sending men and material into orbit.

Numerous designs were considered, such as a fully reusable craft that rode into orbit on the back of another large winged manned booster, but the end result was the now familiar gliding orbiter, external fuel tank and solid rocket booster configuration.

(Above) After the Apollo program ended, NASA decided to concentrate on programs that operated much closer to Earth. (Opposite page, top) President Richard M. Nixon and Dr. James C. Fletcher, NASA administrator, discussed the proposed space shuttle vehicle in San Clemente, California, on Jan. 5, 1972. The president announced that day that the United States should proceed at once with the development of an entirely new type of space transportation system designed to help transform the space frontier into familiar territory. (Opposite page, bottom) These 1975 conception drawings for the space shuttle are very close to the vehicle NASA would ultimately produce.

**Hear Young
and Duke's
comments**
from the lunar
surface

Use your QR Code-Enabled device to
see and hear the sights & sounds of
space shuttle history!

# Developing a
# Design

**E**arly efforts to develop a design for the space shuttle were a direct result of NASA's interest in a permanent space station.

"Everyone was convinced that we had leapfrogged from orbital flight to lunar flight and completely left out a space station," said spacecraft designer Max Faget. "But as we got deeper and deeper into the space station program, it became obvious that we needed a better vehicle. Gemini was too small, and Apollo was too expensive."

During the early shuttle studies, there was plenty of debate over just what the optimal shuttle design — one that best balanced capability, development cost and operational cost — should consist of.

"When we were trying to come up with the space shuttle design," Faget recalled, "I think they had something like about 58 versions as we went through time."

Those versions, as with any major aero-engineering project, benefited greatly from the observation and data gathered from earlier test planes and experimental aircraft. In the early 1960s, NASA test pilots, including Neil Armstrong, flew a rocket-powered aircraft, the X-15, right to the edge of space. Once its initial rocket had burned out, the X-15 essentially became a glider, proving that unpowered descent from altitudes as high as 67 miles was possible.

### See an X-24B lifting body
air-dropped from a NASA B-52

Use your QR Code-Enabled device to see and hear the sights & sounds of space shuttle history!

(Preceding page) NASA did extensive testing of various space shuttle designs at the agency's Ames Research Center. (Above) The M2-F1 and M2-F2 were part of a fleet of lifting bodies flown at the NASA Dryden Flight Research Center that demonstrated the ability of pilots to maneuver and land a wingless vehicle designed to fly back to Earth from space, key research in the development of the space shuttle. (Above right) Dr. Wernher von Braun, a former Marshall Space Flight Center director, and Dr. William Lucas, the newly appointed MSFC director, discuss a model of the shuttle's payload bay in 1974. (Right) A technician at Ames is dwarfed by a 36 percent model.

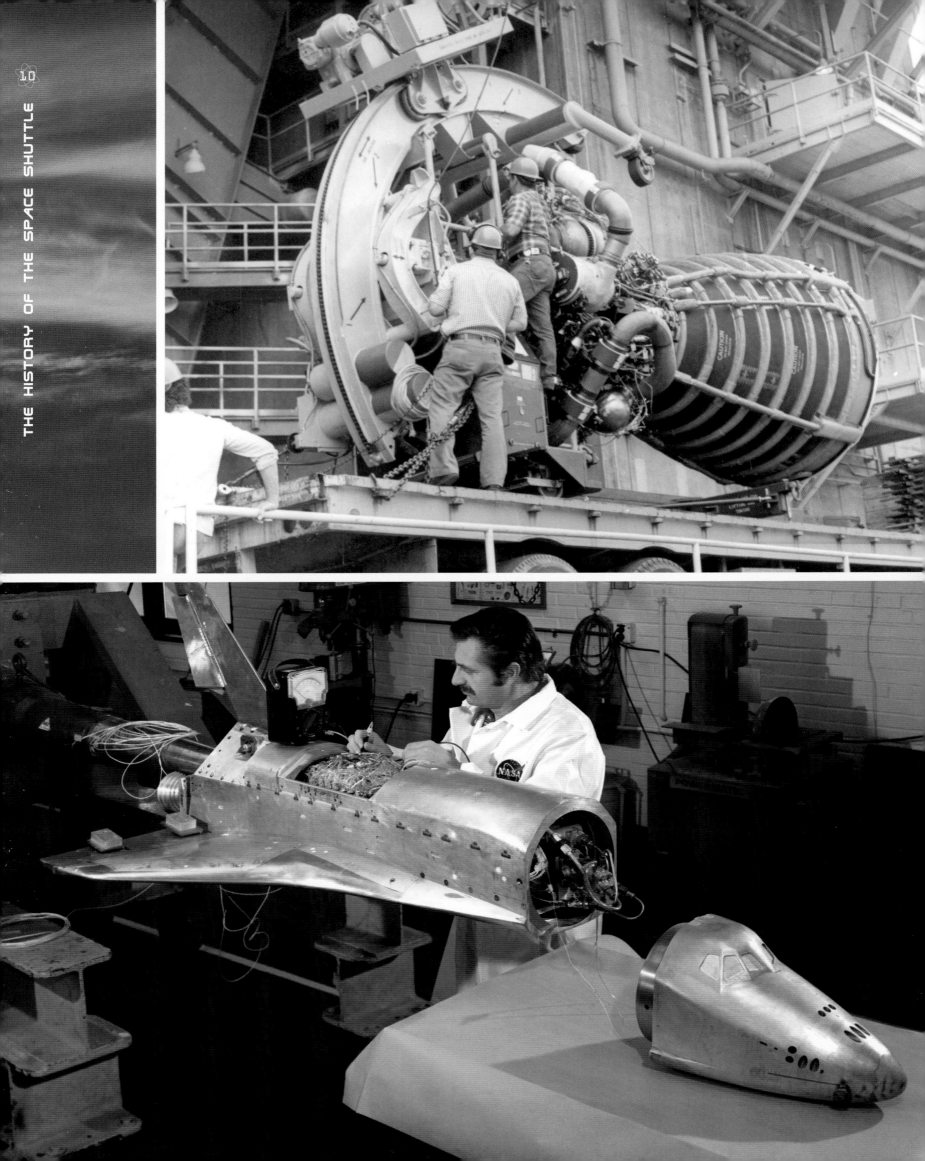

A series of tests on lifting bodies — such as the M2–F2, the HL–10 and the X–24B — proved that pinpoint landings were possible with such craft, which are unpowered and generate lift from the body of the aircraft rather than from wings.

"There needed to be some wing on the shuttle, though," said astronaut and test pilot Joe Engle, "to give it a little better glide ratio, give it a little better controllability and stability than the lifting bodies exhibited."

While the orbiter slowly took shape as the delta wing we know today, other efforts were being directed at the question of just how to place it in orbit. Initially a fully reusable system was preferred in which a large winged booster vehicle would fly the orbiter to a certain altitude, from which the orbiter would propel itself into space. Both vehicles would be able to land and be refurbished for future flights.

## See NASA test pilot
Milt Thompson suit up for test drop of an MF-F2 lifting body

Use your QR Code-Enabled device to see and hear the sights & sounds of space shuttle history!

"When we began to study the thing, we got a better and better idea of the implication of total reusability, and we never did build a completely reusable vehicle," said Faget. "The first thing to go was a reusable first stage. We wanted to make the second stage reusable because we knew we had to get the astronauts down. So we went through a number of steps, but we arrived at the final configuration we know today."

Initially the orbiter was to carry the liquid propellant for its three main engines, but an expendable external fuel tank allowed for a larger payload bay. The fuel tank would be the only piece of the shuttle configuration not used on multiple flights, as the solid rocket boosters were designed to parachute into the ocean for recovery and reuse.

(Preceding page, top) A space shuttle main engine is prepared for a test firing at Stennis Space Center in 1979. (Preceding page, bottom) A technician prepares a 5 percent model of the shuttle for a run in the 16-foot transonic tunnel at NASA's Langley Research Center. (Above) Workmen stack a solid rocket booster in the Dynamic Test Stand at Marshall Space Flight Center before attaching it to an external fuel tank for vibration testing. (Below) The external tanks were produced at the Michoud Assembly Facility in New Orleans. Martin Marietta was the prime contractor that designed and assembled the tanks.

# Testing
## Before The Missions

Once a workable design had been found for the shuttle, production began on the first two orbiters of the program. The second, *Columbia*, would be used for the launch of the first space shuttle mission, STS-1, in 1981, but there was still plenty of training and testing to be done before then.

Most of the testing involving the orbiter would be performed on *Enterprise*, a prototype orbiter that lacked engines and a thermal protection system.

*Enterprise* was originally to be named *Constitution*, but fans from the television show "Star Trek" started a write-in campaign to the White House to ask that the ship be named after Captain Kirk's spacecraft. It worked, and numerous members of the cast were on hand when the vehicle was rolled out of Rockwell's Palmdale assembly facility on Sept. 17, 1976.

*Enterprise* was used for NASA's Approach and Landing Test program, making 13 flights, eight manned, attached to a modified 747, the Shuttle Carrier Aircraft.

### See *Enterprise* separate from
the Shuttle Carrier Aircraft during testing

Use your QR Code-Enabled device to see and hear the sights & sounds of space shuttle history!

"When we first rode on top [of the SCA], you couldn't see the 747, no matter how you'd lean over and try to look out the side window, and you couldn't view any part of it," said astronaut Fred Haise, who commanded the shuttle's first captive flight. "So it was kind of like a magic carpet ride, you know. You're just moving along the ground and then you take off."

(Opposite page, top) Fans of "Star Trek" conducted a write-in campaign to have NASA name the first shuttle after Captain Kirk's *USS Enterprise*, and the cast was there to celebrate the rollout. (Above left) Although *Enterprise* would only fly approach and landing tests, five of those flights would be manned by Fred Haise and C. Gordon Fullerton. (Left) Shortly after rollout, *Enterprise* was ferried to Marshall Space Flight Center and lifted into the center's Dynamic Test Stand for vibration testing. (Opposite page, bottom) Built in 1964 to conduct tests on the Apollo-era Saturn V rocket, the 360-foot-high test stand was modified to accept the new space shuttle.

During the last five test flights, *Enterprise* separated from the jet at altitude and was then piloted by two astronauts to a landing at Edwards Air Force Base. For the captive flights and the first three free flights, an aerodynamic fairing covered the orbiter's aft end. Three dummy main engines were installed for the final two flights to simulate weight and aerodynamic characteristics of an operational orbiter. The tests were designed to place the vehicle in aerodynamic flight by itself and exercise all of the systems: hydraulic, electronic, flight control and landing gear.

### See *Enterprise* land at Edwards AFB
during testing

Use your QR Code-Enabled device to see and hear the sights & sounds of space shuttle history!

"For ALT, we were clearly test pilots because we were doing stuff that there wasn't a procedure for," said astronaut C. Gordon Fullerton, co-pilot for five of the eight manned flights. "We were writing the procedure and then flying it for the first time."

After those tests, *Enterprise* was flown to NASA's Marshall Space Flight Center in Huntsville, Alabama, where it was mated with the external tank and solid rocket boosters, and subjected to a series of vertical ground vibration tests. The orbiter was also sent to the Kennedy Space Center in Florida, where it was rolled out to the launch pad to act as a "stand-in" as NASA prepared for the first shuttle launch.

(Preceding page, top and left) For several of the vibration test configurations, *Enterprise* was mated with an external fuel tank and two solid rocket boosters filled with different levels of inert propellants to simulate different parts of a flight. (Preceding page, right) NASA conducted drop tests of solid rocket booster articles to validate the boosters' recovery system. (Right) *Enterprise* was used in a series of approach and landing tests, first attached to the Shuttle Carrier Aircraft and then in free flight, where Haise and Fullerton guided the orbiter to several landings at Edwards Air Force Base.

### Hear
### John Young
### talk about

the future of

the shuttle

Use your QR Code-Enabled device to see and hear
the sights & sounds of space shuttle history!

# THE FIRST SHUTTLE LAUNCH

The first launch of America's space shuttle, mission STS-1, has been called "the boldest test flight in history." The basic objectives were to demonstrate the safe launch into orbit and return to landing of *Columbia* and its crew, and verify the combined performance of the entire shuttle vehicle — orbiter, solid rocket boosters, external tank — up through separation and retrieval of the spent solid rocket boosters.

Never before had the agency launched a manned spacecraft without a prior unmanned test flight, but NASA was confident in mission Commander John Young, a veteran of four missions in three types of spacecraft.

"If you want to go into space for the first time on a new vehicle that's never been flown, you want to go with a pro," said Young's co-pilot, Robert Crippen, "and John certainly is a pro."

Crippen was making his first spaceflight, but his mastery of *Columbia*'s sophisticated computer systems garnered Young's appreciation.

"I was really lucky to have Bob Crippen with me because he knew all the software end to end," said Young. "He was a swell fellow and really smart about the vehicle."

From its liftoff at Cape Kennedy to its successful landing at Edwards Air Force Base, STS-1 proved to be a stunning success. The shuttle's worthiness as a space vehicle was verified, and the data and observations gathered from *Columbia*'s two days in space advanced NASA's knowledge of shuttle space flight by a thousand fold.

As Johnson Space Center Director Chris Kraft said, "We just got infinitely smarter."

(Preceding page, top) NASA chose veteran astronaut John Young, left, to command the shuttle's maiden flight and paired him with rookie astronaut Robert Crippen. (Preceding page, bottom left) The shuttle *Columbia*, the first fully operational orbiter, rolls out of the assembly building on the way to the launch pad. (Above) The external tank was painted white for the shuttle's first mission, a practice that NASA would shortly abandon. (Preceding page, right) The first shuttle flight was a huge success, but the orbiter's journey wasn't truly over until Shuttle Carrier Aircraft pilot Fitz Fulton, second from right, ferried *Columbia* from its landing site at Edwards AFB back to Kennedy Space Center.

# Intense
# **Training**

**M**ission training for the first shuttle launch was akin to the first Mercury flights — no one had ever done this before, and the trainers and the simulation supervisors were learning right alongside the astronauts.

"Prior to doing the first flight, all we had to go on was the engineering data that we had, and we used that to make sure of what we could do, as far as you could reach something or you couldn't reach something during ascent or entry," said STS-1 co-pilot Robert Crippen. "So, there are certain switches that weren't accessible during those periods of time, so we had to make sure that the procedures would match what our capabilities were."

The training was intense, and the mission simulator operators threw everything they could think of at the astronauts.

**See video
from the cockpit**
of the Shuttle Training Aircraft

Use your QR Code-Enabled device to see and hear the sights & sounds of space shuttle history!

"What the simulation supervisors were primarily trying to do is to make sure that the procedures we had in place would work and that we knew how to use them," said Crippen. "That was for both us in the cockpit and for the Mission Control folks, because we all had to work together. They would run you through the wringer, which is what we wanted them to do. If there was something that could fail, we wanted to make sure we could deal with it."

Not only was training intense, it was practically round the clock. During a press conference two weeks before the launch, John Young went over the crew's training schedule for just the following week. It was packed with integrated entry simulations, integrated abort simulations, the first ascent launch readiness verification test in *Columbia*, and runs in the shuttle

(Left) Robert L. Crippen, pilot for the first space shuttle orbital flight test, descends steps leading into the water immersion tank in the Johnson Space Center's training and test facility. (Opposite page) STS-1 Commander John Young goes through a spacesuit-donning exercise during a brief period of microgravity afforded onboard a KC-135 flying a parabolic curve.

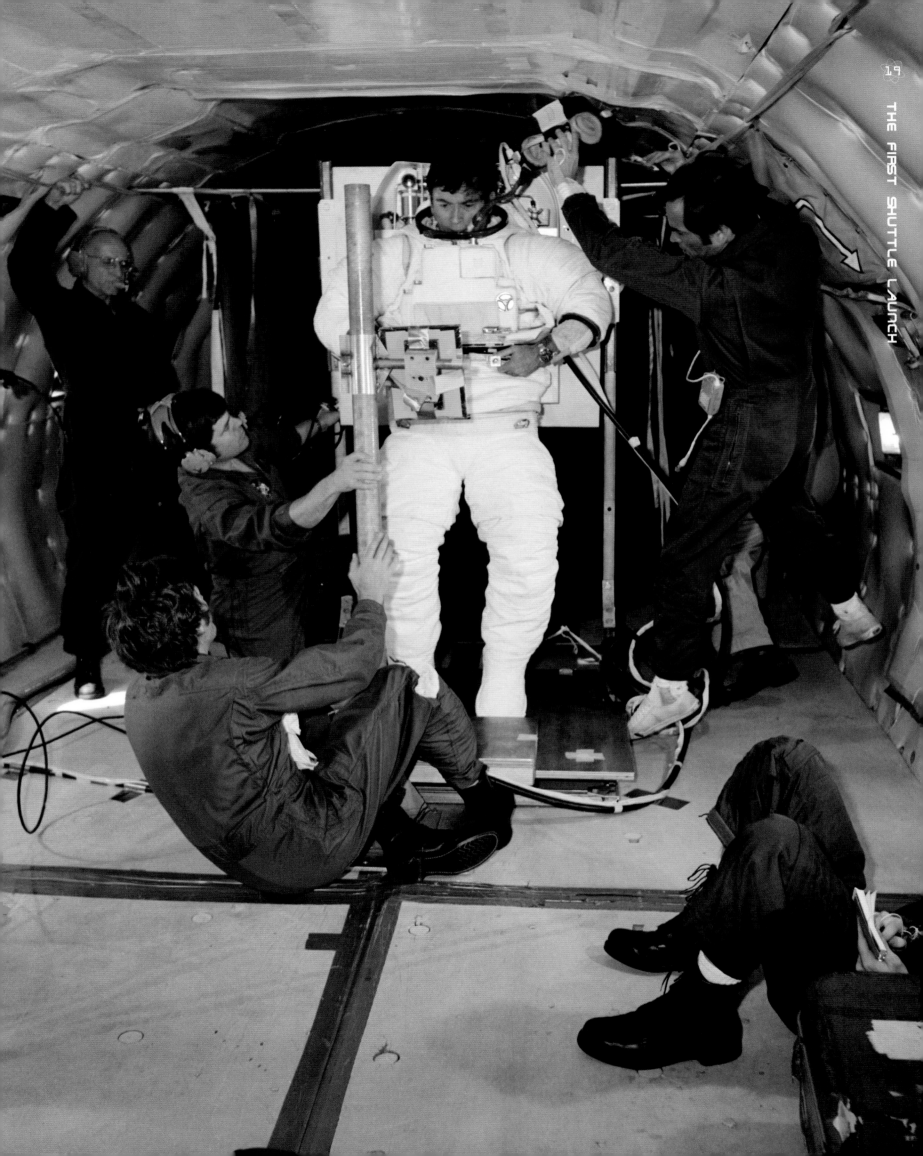

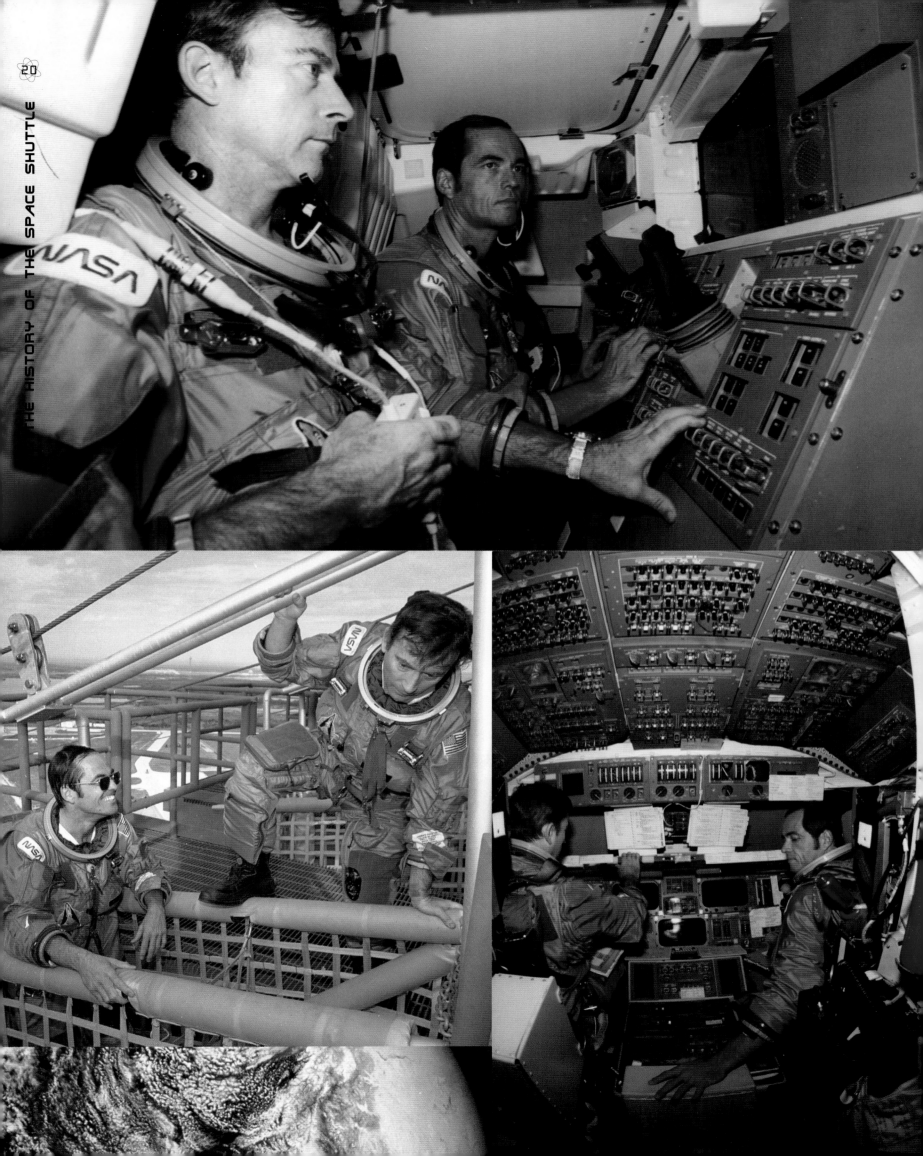

training airplane, a highly modified Gulfstream-2 aircraft that mimicked the shuttle's unique flying characteristics.

"The folks did a marvelous job coming up with that vehicle," said Crippen. "The ground-based simulators give you some idea of what it's like to fly, but it's not like the real thing, and the STA does an excellent job of that. We did extensive approaches, primarily using the runway out in White Sands, New Mexico, but we also flew at Edwards, and we flew it at the runway at the Kennedy Space Center. I think John and I had on the order of 1,500 approaches each before we flew that first flight."

  **Hear Vice President Bush's call** to Columbia

Use your QR Code-Enabled device to see and hear the sights & sounds of space shuttle history!

Things weren't all hard work and no play, though. Less than a month before the historic launch, Young and Crippen received a visit from Vice President George H.W. Bush at Kennedy Space Center.

"We had him up in the cockpit of *Columbia* and looked around; went out jogging a few miles with him," said Crippen. "So we felt like we had a personal rapport with him, and so when we got a call [during the mission] from the vice president, it was like talking to an old friend."

That was the goal of all the STS-1 mission training. To make spacecraft and the procedures needed to fly it so familiar to the astronauts that the launch itself was as comfortable as two friends talking.

(Preceding page, bottom left) Robert Crippen smiles as John Young boards the emergency pad escape system known as the "slidewire" baskets during a terminal countdown demonstration test at Kennedy Space Center. (Preceding page, bottom right) Young and Crippen log time in the shuttle orbiter *Columbia* in the Orbiter Processing Facility at the Kennedy Space Center. (Right) Vice President George H.W. Bush visited KSC in March of '81, after President Ronald Reagan was forced to cancel his visit due to the attempt on his life just a few weeks earlier.

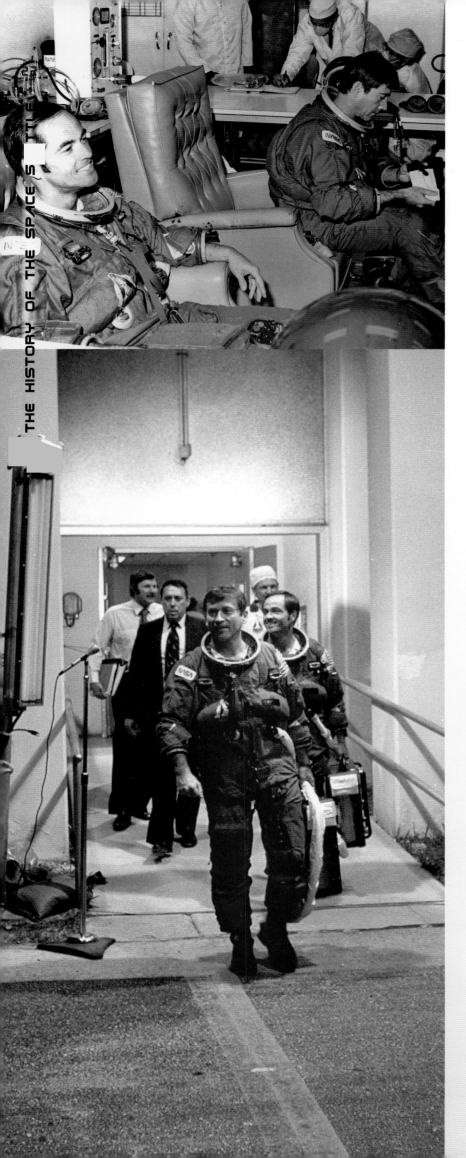

# Launch
## Into The Future

STS-1 was initially set to launch April 10, 1981, but the flight had to be scrubbed after engineers discovered a problem with the spacecraft's computer system. The shuttle's four primary computers weren't communicating with the backup computer, something the engineers, and the astronauts, considered essential to a safe flight.

"The vehicle is so complicated, I fully anticipated that we would go through many, many countdowns before we ever got off," said Robert Crippen. "But I was really surprised, because that was the area I was supposed to know, and I had never seen this happen; never heard of it happening."

**Hear Launch Director George Page read** a message from President Ronald Reagan to the STS-1 astronauts

Use your QR Code-Enabled device to see and hear the sights & sounds of space shuttle history!

With so much anticipation for the shuttle's first flight, NASA had every available resource working to find a solution to the problem. It ended up being a simple fix, a matter of initializing the backup computer at the right time, and by April 12, the launch was back on. So John Young and Crippen again climbed into *Columbia*, ready to blast off into the unknown of shuttle spaceflight.

"In talking prior to going out there, Launch Director George Page told John and I, he says, 'Hey, I want to make sure everybody's really doing the right thing and focused going into flight. So I may end up putting a hold in that is not required, but just to get everybody calmed down and making sure that they're focused.'"

Page called for the hold around T–9:00 and took advantage of the break to read a message to the astronauts from President Ronald Reagan. Reagan was scheduled to visit

---

(Above left) Robert Crippen smiles as John Young goes over some last-minute details in the suit-up room at Kennedy Space Center. (Opposite page, top) Technicians check the fit of the two astronauts' launch and re-entry suits. (Opposite page, bottom) As the launch hour nears, Crippen gets down to business, using his hands to illustrate some technical point to Young. (Left) Young and Crippen head out to the launch pad to climb into *Columbia*, the first time in NASA history that the first launch of a new vehicle would be manned.

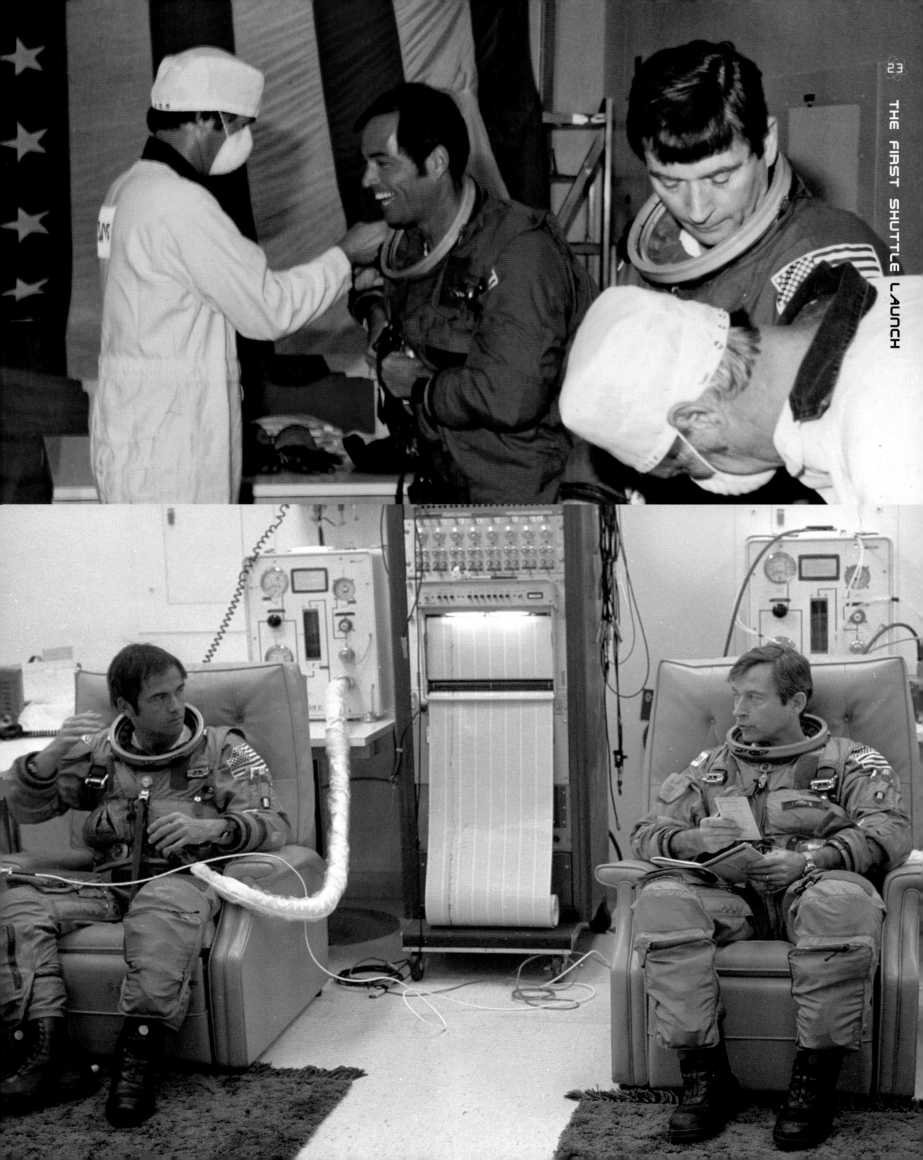

Mission Control during STS-1 but cancelled because he was still recovering from the assassination attempt by John Hinckley Jr. on March 30.

As the clocked ticked down to America's first space shuttle launch, the astronauts began to get a sense that today might really be the day.

"About one minute to go, I turned to John. I said, 'I think we might really do it,' and about that time, my heart rate started to go up," said Crippen. "I think they said it was up to about 130. John's was down about 90. He said he was just too old for his to go any faster."

At precisely three seconds after 7 a.m. local time, *Columbia* roared from the launch pad. The shuttle's three main engines were pouring out almost 1.2 million pounds of thrust while the SRBs added 6.6 million more.

"There was no doubt you were headed someplace," said Crippen. "It was a nice kick in the pants."

## Watch a video of the STS-1 launch

Use your QR Code-Enabled device to see and hear the sights & sounds of space shuttle history!

Just six seconds after liftoff, STS-1 cleared the launch tower and two seconds later began its pitchover maneuver. Everything was going just as planned. The SRBs jettisoned a little over two minutes into the flight and the external tank fell away a few minutes later. All in all the shuttle had accelerated from zero to a speed of almost 18,000 miles per hour, nine times as fast as the average rifle bullet, before it finally reached orbit.

"We were delighted when we got into orbit," said Young. "We learned that we can build a complicated vehicle and make it work very well."

(Preceding page inset) Spotlights illuminate *Columbia* as she sits at Launch Pad A, Complex 39, waiting for Young and Crippen. (Preceding page) After a planned hold at around T-minus 9:00, *Columbia* lifted off into the early morning sky. (Above right) Thousands of people watched as the shuttle lifted off, hoping that all the work they had done to build this new space vehicle would be successful. (Right) The solid rocket boosters fell away just as planned, and a few minutes later *Columbia* was safely in orbit.

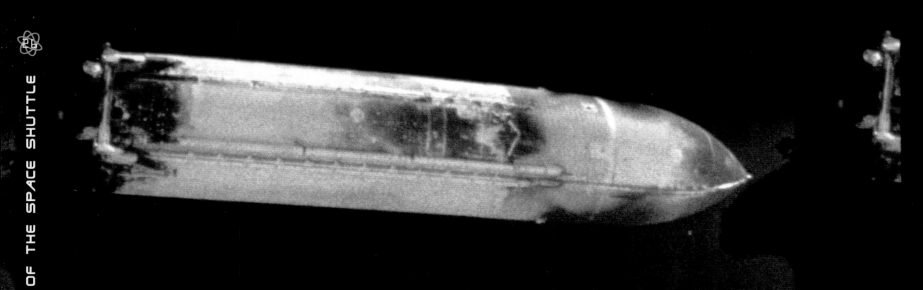

# Camping Out In Orbit

**J**ust hours after John Young and Robert Crippen blasted off of Pad 39A at Kennedy Space Center, *Columbia* was given the go ahead for orbit, clearing the shuttle to stay in space for its full two-day mission.

STS-1's main objective was to test the vehicle's performance as a reusable spacecraft using two equipment packages within the shuttle's payload bay. The Developmental Flight Instrumentation and the Aerodynamic Coefficient Identifications Package pallet recorded temperatures, pressures, acceleration levels and other forces on the craft throughout the flight.

Another key test involved the successful operation of the massive payload bay doors, which are essential to the shuttle's cargo capability. The astronauts had successfully opened and closed both doors before *Columbia* achieved its final orbital altitude and reopened the payload bay for the rest of the flight. The successful operation of the doors provided a clear view of the craft's Orbital Maneuvering System (OMS) pods, which showed signs of heat-shield tile damage.

"At that time I saw, back on the Orbital Maneuvering System's pods, that there were some squares back there where obviously the heat shielding tiles were gone," said Crippen. "They were dark instead of being white."

Houston was worried that tiles might also be missing on the bottom of the orbiter, which would take the brunt of the massive heating of re-entry, but Crippen wasn't concerned. "I knew that all the critical tiles, the ones primarily on the bottom, we'd gone through and done a pull test with a little device to make sure that they were snugly adhered to the vehicle."

While there were other, smaller mishaps, such as the shuttle's toilet not working properly, both astronauts found

**Hear Houston give Columbia a "go" to stay on orbit**

Use your QR Code-Enabled device to see and hear the sights & sounds of space shuttle history!

(Preceding page, top) *Columbia*'s external tank falls away to burn up in the atmosphere as the shuttle reached its planned orbit. (Preceding page, center) John Young logs flight data in a loose-leaf flight activities notebook onboard from the commander's station on *Columbia*'s forward flight deck. (Preceding page, bottom) Young appears on screen in Mission Control during the flight. (Above right) Robert Crippen floats near the spacecraft's top viewing windows. (Right) As part of their mission duties, Crippen and Young took a series of photos of the Earth, including this one of Eleuthera Island in the Bahamas and part of the Great Bahama Bank.

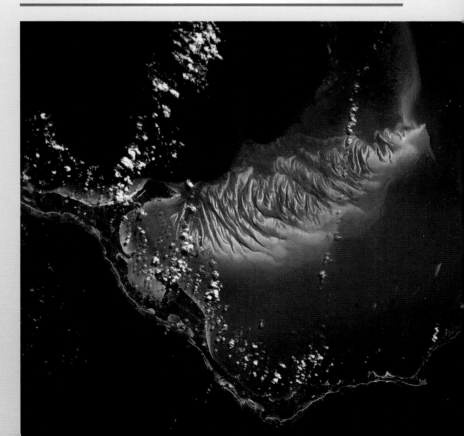

🔊 **Hear Commander John Young describe flying** over Kennedy Space Center

Use your QR Code-Enabled device to see and hear the sights & sounds of space shuttle history!

time to enjoy the ride, peering out the windows as the Earth passed by below and snapping photos of each other in zero G. Crippen especially was soaking in the unique sensation.

"The vehicle was big enough that you could move around quite a bit," he said. "You could take advantage of being weightless, and it was delightful. It was a truly unique experience, learning to move around."

Living in the shuttle was little bit like camping out, he said. Even the food wasn't half bad, a serious improvement from the Gemini and Apollo days for sure.

"We even had steak," Crippen recalled. "It was irradiated so that you could set it on the shelf for a couple of years, open it up, and it was just like it had come off the grill. It was great."

Crippen even found time to do some international outreach, playing a tape of "Waltzing Matilda," Australia's unofficial national anthem, as the shuttle passed over the Aussie ground stations, "hoping that maybe somebody would give us an invite to go to Australia, since I'd never been there."

(Left) Crippen prepares a meal aboard *Columbia*, one of the first of the shuttle era. (Above center) Crippen does a cannonball in the middeck area of *Columbia*. (Above left) Young and Crippen discovered that several of the heat-resistant tiles covering the shuttle's engine cowlings were missing, but "that didn't cause me any great concern," said Crippen, "because I knew that all the critical tiles, the ones primarily on the bottom, we'd gone through and done a pull test with a little device to make sure that they were snugly adhered to the vehicle." (Opposite page, top) Mission Control monitors activities in the later part of the flight. (Opposite page, bottom) Young found time in his schedule to shave on orbit.

# A Historic
# Landing

On the morning of April 14, 1981, the space shuttle *Columbia* fired up its engines for a de-orbit burn that would send the vehicle home after two days in space.

"We did our de-orbit burn on the dark side of the Earth and started falling into the Earth's atmosphere," said co-pilot Robert Crippen. "It was still dark when we started to pick up outside the window; it turned this pretty color of pink. It was just a bunch of little angry ions out there that were proving that it was kind of warm outside, on the order of 3,000 degrees. It was kind of like you were flying down a neon tube."

That plasma field surrounding the shuttle cut off communications with the ground temporarily, a situation the astronauts call LOS, loss of signal.

"They go into blackout," said astronaut Fred Hauk, who was in Mission Control at the time. "And you're just waiting there and you're hoping you'll hear them start talking when they're supposed to. It was probably within a matter of 10 or 20 seconds from the time when we should hear them that we did hear them, and we thought, 'Whew. They've survived so far.'"

The shuttle was still performing beautifully as the autopilot began a series of S-turns. Commander John Young took over manual control at around 40,000 feet and began to steer the orbiter over the San Joaquin Valley, headed for Edwards Air Force Base. "What a way to come to California," Crippen remarked over the radio.

NASA chase planes followed *Columbia* down to her final approach and landing on the runway, calling out distances from the shuttle to the ground for Young in the cockpit.

"My primary job was to get the landing gear down," Crippen remembered, "which I did, and John did a beautiful job of touching down."

Masses of people from up and down the West Coast had started lining up at the base's gates the day before the landing, and by the time the shuttle touched down, well over 200,000 people were in attendance.

**Hear audio from the shuttle landing**

Use your QR Code-Enabled device to see and hear the sights & sounds of space shuttle history!

(Preceding page, bottom left) Lakebed Runway 23 at Rodgers Dry Lake adjacent to Edwards Air Force Base served as the landing site for the first space shuttle mission. (Preceding page, top and bottom right) The shuttle has no engines designed for maneuvering upon landing; it essentially becomes a giant glider upon re-entry. (Top right) The shuttle slows from 17,000 mph, its speed while orbiting the Earth, to about 215 mph at the moment of touchdown. (Right) Thousands gathered to see the historic moment.

**See John Young exit**
the shuttle after landing

Use your QR Code-Enabled device to see and hear the sights & sounds of space shuttle history!

"I'll never forget it," said James Young, chief historian of the Air Force Flight Test Center. "You just had to be there to hear, even feel, the double crack of the sonic boom. It was such a tremendous sense of excitement to see something never seen before, to witness such a historic event."

Even John Young, the stoic veteran of both the Gemini and Apollo programs, was caught up in the excitement, joking over the radio that he ought to just pull the shuttle right on up into the hanger. As the shuttle came to a halt, Young couldn't wait to get the door open.

"He couldn't sit still, and I thought he was going to open up the hatch before the ground did," said Crippen. "But they finally opened up the hatch, and John popped out. Meanwhile I'm still up there, doing my job, but I will never forget how excited John was."

(Above) Workers in Mission Control in Houston watch video of the shuttle's landing in California. (Left) Technicians congratulate Young and Crippen on their successful flight. (Opposite page, bottom left) Shuttle support vehicles approach *Columbia* after landing. (Opposite page, top right) The purge and coolant umbilical access vehicles pump coolant and purge air through the umbilical lines. Purge air provides cool and humidified air conditioning to the payload bay and other cavities to remove any residual explosive or toxic fumes that may be present. (Opposite page, center right) After the shuttle has been cleared, it is towed to the Mate Demate Device to be placed atop the Shuttle Carrier Aircraft (opposite page, bottom right) for transport back to Kennedy Space Center.

**See a short video
of some of the
space shuttle's
many firsts**

# SHUTTLE FIRSTS

**W**hen NASA first sent men into space in the 1960s, the original astronauts were chosen from military test pilots, like Alan Shepard and John Glenn, who were wedged into one-man Mercury and two-man Gemini capsules. The Apollo missions to the moon required three men, as two went to the surface of the moon, while one other orbited above.

The arrival of the larger-sized space shuttle in 1981 meant that NASA could have a wider range of specialized candidates, many of whom who did not have pilot training or did not serve in the armed forces. Crew sizes increased

The frequency of the shuttle's planned flights and the changing social landscape in America made it possible for shuttle crews to record a significant number of space "firsts." (Preceding page) The crew of STS-7 released the SPAS satellite, which took the first-ever pictures of the shuttle in orbit. (Above) STS-61A was the first time that eight astronauts were sent into orbit at the same time in the same vehicle.

as needs were identified by NASA, and mission specialists, including scientists, were added. The lure of the astronaut program also attracted a diverse pool of candidates, including women and people of various races and nationalities, all of whom made history when they flew aboard the shuttle.

The entire design of the reusable shuttle — from the solid rocket boosters, to the large cargo bay in the center of the spacecraft that contained satellites and experiments inside, to the wings that allowed for a gliding return to Earth, even at night — gave birth to several firsts: untethered space walks; the use of a robotic arm; satellites collected and released into orbit; deployment of a deep-space probe; docking of the shuttle to the Russian Mir space station; and its work in assembling the International Space Station.

Almost every shuttle mission had a first of some type, from personnel to payloads, over its 30 years of flight.

# How Many People Can This Thing Fit?

**W**hen the space shuttle first headed to space in 1981, there were just two men aboard the orbiter *Columbia*: Commander John Young, a veteran of two Gemini and two Apollo space flights, and pilot Robert Crippen, a U.S. Navy captain who had never before flown in space. That first flight carried no payloads, as the shuttle systems were being tested.

The first four missions were engineering test flights, and all had two-person crews. When the shuttle was fully operational for STS-5 in November 1982, *Columbia* carried four men: Commander Vance D. Brand, pilot Robert F. Overmyer, and mission specialists Joseph Allen and William Lenoir. It was the first four-person crew in a spacecraft for NASA.

The crew total moved to five in June 1983 during STS-7 when *Challenger* carried Commander Crippen, pilot Frederick Hauck, and mission specialists John Fabian, Sally Ride and Norman Thagard. Ride also made history during the mission as the first American female astronaut in space.

Two missions later on STS-9, six people — Commander John Young, pilot Brewster Shaw Jr., mission specialists Owen Garriott and Robert Parker, and payload specialists Byron Lichtenberg and Ulf Merbold of the European Space Agency — rode to space on *Columbia* in November 1983.

The total number of crewmembers fluctuated between five and six for the next three missions before the shuttle set another record with seven aboard *Challenger* in October 1984 for STS-41G: Commander Crippen, pilot Jon McBride, mission specialists Ride, Kathryn Sullivan and David Leestma, and payload specialists Marc Garneau and Paul Scully-Power. It was the first flight with two women.

(Top left) The shuttle's first four flights, checkout missions for the hardware, only carried two crew members each. STS-5, the shuttle's first operational flight, carried four astronauts, the first time that four Americans had launched into space on the same vehicle. (Left) Vance Brand, Robert Overmyer, Joe Allen and William Lenoir were tabbed to be NASA's first four-man crew. (Opposite page, top left) The shuttle was able to accommodate four astronauts in seats on the flight deck, but as crews grew larger, it was necessary to install other seats, such as the one Bill Shepherd and Guy Gardner are clowning around with, on the middeck. (Opposite page, top right) The seats are broken down and stored while in orbit. (Opposite page, middle right) Crew size expanded all the way to seven on STS-41G and to a record eight on STS-61A (opposite page, bottom).

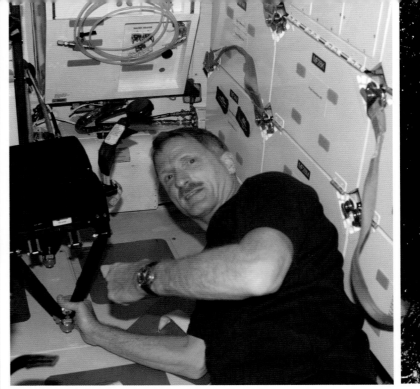

Either five or seven crewmembers flew on the next eight missions, until a shuttle record of eight was set in October 1985. STS-61A carried Commander Henry Hartsfield Jr., pilot Steven Nagel, mission specialists Guion Bluford Jr., James Buchli and Bonnie Dunbar, and payload specialists Reinhard Furrer, Ernst Messerschmid and Wubbo Ockels.

## Hear audio from the landing of STS-61A,

which carried the largest crew in history up to that time

Use your QR Code-Enabled device to see and hear the sights & sounds of space shuttle history!

Seven flew on the next three missions, including the ill-fated *Challenger* on Jan. 28, 1986, aboard STS-51L. After the tragedy, crew size was reduced to five when shuttle flights resumed in September 1988.

Twelve missions carried five astronauts before seven departed on *Columbia* for STS-35 in December 1990.

The record of eight astronauts was equaled for the only time in June 1995 on STS-71, which was the first shuttle-Russian Space Station Mir docking mission. Seven astronauts left on *Atlantis* for Mir: Commander Robert Gibson, pilot Charles Precourt, mission specialists Ellen Baker, Bonnie Dunbar and Gregory Harbaugh, along with the uploading Mir 19 crew of Anatoly Solovyev and Nikolai Budarin. During the joint mission operations, the Mir 19 crew took over the Russian station, and the Mir 18 crew of Thagard, Vladimir Dezhurov and Gannady Strekalov boarded *Atlantis* for the ride home.

Between five and seven crewmembers flew subsequent missions, including Mir dockings and the work on the ISS.

When the shuttle program came to an end with STS-135 on July 8, 2011, there were four crewmembers onboard *Atlantis*, the lowest total since the sixth flight of the shuttle fleet in April 1983: Commander Chris Ferguson, pilot Doug Hurley, and mission specialists Sandy Magnus and Rex Walheim.

(Preceding page, top) STS-61A was a Spacelab mission funded by West Germany that included several German mission specialists. The crew slept in two shifts to ensure that the lab was being manned around the clock. (Preceding page, bottom) STS-71 nearly equaled the total of the Spacelab mission, launching with seven astronauts aboard. The shuttle docked with the Russian Mir space station and came home with three of Mir's crew, for a total of eight astronauts on the return flight. (Above right) The crew of STS-61A, shown here departing the shuttle at the close of the mission, still holds the record with eight, however. (Right) Recent shuttle flights have nearly all had five or more crewmembers, necessitating the middeck seats at launch and re-entry.

# Ride, Sally Ride

The U.S. manned spaceflight program was an all-male affair through the Mercury, Gemini and Apollo missions. But that would change when the space shuttle program was started, and women would play a large part in the 30-year history of shuttle missions.

The first six shuttle missions, including the four research and development flights, had all-male crews. STS-7, which launched June 18, 1983, would have the first American female astronaut aboard — Sally Ride.

Ride, after answering an ad in the Stanford University student newspaper, was selected as an astronaut candidate in January 1978, and after she completed a one-year training program, she was eligible for mission specialist duty. She served as the capsule communicator for two of the early shuttle flights before her mission on the orbiter *Challenger*.

## Hear Sally Ride compare her first shuttle launch

🔊 to a Disneyland E-ticket, which admitted the bearer to the newest and most advanced rides and attractions

Use your QR Code-Enabled device to see and hear the sights & sounds of space shuttle history!

Prior to her mission, she worked with engineers on testing and refinement of the robotic arm that would be used to release and catch satellites in orbit over the life of the shuttle fleet. She then operated the robot arm during STS-7.

"It was a little easier to use the arm in space than it was in the simulators, because I could look out the window and see a real (robot) arm. And although the visuals in the simulators are very good, there's nothing quite like being able to look out the window and see the real thing. It felt very comfortable and familiar."

(Top left) Sally Ride, shown here speaking to the Cleveland Women's City Club in 1983, was one of NASA's first female astronaut candidates (center left). (Left) Ride would ultimately be chosen to become the first American woman in space, launching aboard the *Challenger* on STS-7. (Opposite page, top) Six missions later, Kathryn Sullivan joined Ride on STS-41G, the first mission ever to send two women into space together. (Opposite page, bottom) Sullivan, shown here with Ride and a sleep-restraint device, performed a spacewalk during the mission, becoming the first American woman to do so.

In October 1984, Kathryn Sullivan joined Ride on STS-41G for the first spaceflight to include two women. Sullivan also made history during that mission as the first American woman to walk in space, which she did for three-and-a-half hours.

"I could have been the 50,000th or 100,000th woman or human being to do a spacewalk," Sullivan said. "In terms of the broad historical backdrop, it still would have been my first spacewalk. I honestly can't think of anything I would have done differently in how I went about it, how I prepared for it, how I was processing the meaning of it to me and to the mission, and to everything else."

STS-47 in September 1992 had the first African-American female astronaut, Mae Jemison, onboard the orbiter *Endeavour* as the science mission specialist.

The first Hispanic female astronaut was Dr. Ellen Ochoa, who flew on *Discovery* during the STS-56 mission in April 1993. Ochoa later became the deputy director of the Johnson Space Center.

 **Watch a clip of First Lady Hillary Clinton**
announcing Eileen Collins as the first female shuttle commander

Use your QR Code-Enabled device to see and hear the sights & sounds of space shuttle history!

Eileen Collins was the first female astronaut to be a shuttle pilot (STS-63, February 1995) and a shuttle commander (STS-93, July 1999). She was also commander of the STS-114 "Return to Flight" mission in July 2005, the first after the 2003 *Columbia* tragedy.

When asked to reflect prior to STS-114, Collins said, "My heroes were the astronauts that have gone before me, the test pilots and women pilots who flew back in World War II and the women who went through the medical testing for the Mercury program. All of these people that I read about as I grew up through high school and college have influenced me in a positive way."

(Preceding page, bottom and top left) The first African-American woman in space, Mae Jemison, served as a mission specialist on STS-47, a joint mission between NASA and the National Space Development Agency of Japan, which conducted microgravity investigations in materials and life sciences. (Preceding page, top right) Dr. Ellen Ochoa became the first female Hispanic astronaut when she lifted off on STS-56 in 1993 as a mission specialist. (Above right) Ochoa would go on to fly three more missions and currently serves as the deputy director of Johnson Space Center. (Center right and right) Eileen Collins served as the first female pilot of the space shuttle on her first mission, STS-63. Collins would also become the first female commander of the shuttle on STS-93.

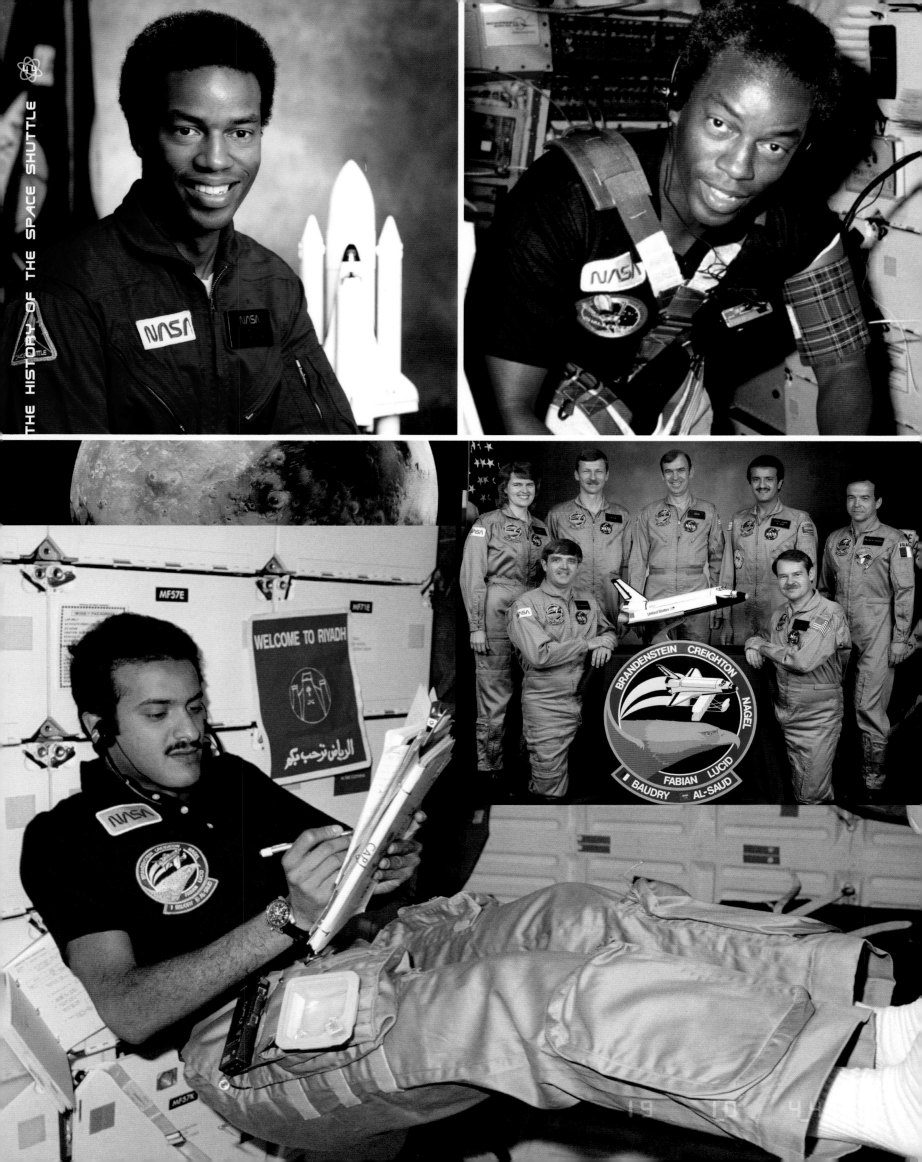

WELCOME TO RIYADH

الرياض ترحب بكم

# The Whole World Is Our Home

The diversity of NASA's expanded astronaut corps for the space shuttle program would be on display during the 135 missions flown over 30 years.

The first African-American male to be a part of a shuttle mission was Guion Bluford Jr., who joined the astronaut group in August 1979. The Penn State graduate was part of the Air Force ROTC in college, and he became a pilot who flew 144 combat missions in the Vietnam War.

His first shuttle flight was aboard STS-8 in August 1983 as a mission specialist on the orbiter *Challenger*, the first mission with a night launch and night landing.

Bluford served on three other missions: STS-61A in 1985, STS-39 in 1991 and STS-53 in 1992. In all, he logged over 688 hours in space.

When asked in 2004 about his impact on the space program, he said, "I was very proud to have served in the astronaut program and to have participated on four very successful space shuttle flights. I also felt very privileged to have been a role model for many youngsters, including African-American kids, who aspired to be scientists, engineers and astronauts in this country."

International astronauts were part of the shuttle program as payload specialists, starting in the mid-1980s, including the first Arabian to fly in space on the shuttle, Sultan Salman Abdulaziz Al-Saud, a payload specialist on *Discovery* in June 1985 during the STS-51G mission.

Franklin Chang-Diaz was the first Hispanic U.S. astronaut, and first naturalized U.S. citizen, to go into space. The Costa Rican-born Chang-Diaz flew first on STS-61C in 1986, and he is tied for the most shuttle missions by one astronaut with seven.

(Preceding page, top left and right) Guion "Guy" Bluford Jr., chosen in the same astronaut class as Sally Ride, followed Ride into history by becoming the first African-American to fly into space, on the mission after Ride's, aboard STS-8. (Preceding page, middle right) The crew of STS-51G included Saudi Arabia's Sultan Salman Al Saud, who was on board to oversee the deployment of Arabsat. (Preceding page, bottom) Al Saud became the first Arab, the first Muslim and the first member of a royal family to fly into space. (Right) Franklin Chang-Diaz, who also shares the record for the most spaceflights with Jerry Ross, became the first Hispanic astronaut to fly on an American space flight on STS-61C in 1986.

**Hear President Ronald Reagan congratulate Guion Bluford** on becoming the first African-American in space

Use your QR Code-Enabled device to see and hear the sights & sounds of space shuttle history!

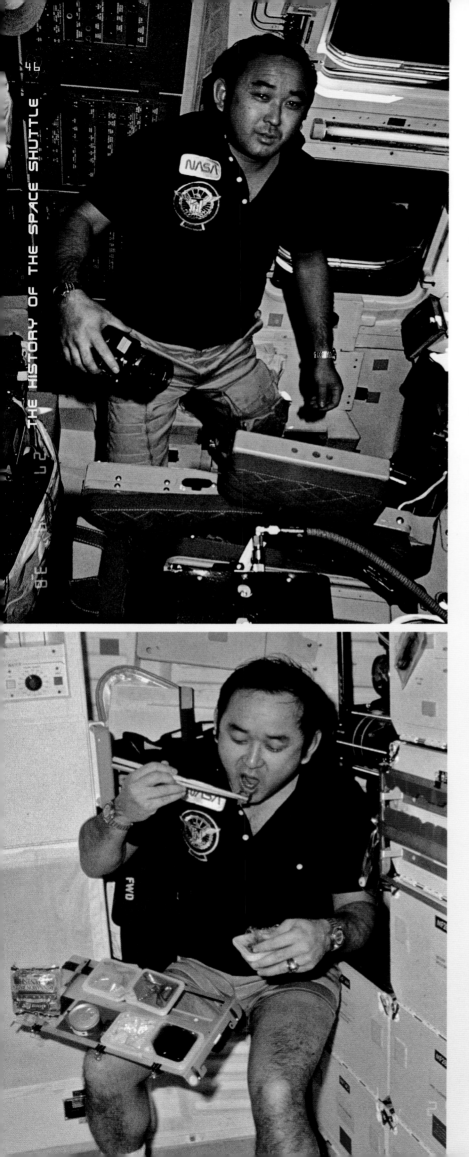

An African-American commanded the shuttle for the first time when Frederick Gregory led STS-33 on *Discovery* in November 1989. Gregory, an Air Force Academy graduate who flew in Vietnam, was selected as an astronaut in 1978. He was a pilot on STS-51B, and served as commander twice, as he was also in charge of STS-44 in November 1991. Gregory later served as a deputy administrator for NASA.

## Hear astronaut Mike Mullane read a message in Arabic

from the ground to Sultan Salman Abdulaziz Al-Saud in orbit aboard STS-51G

Use your QR Code-Enabled device to see and hear the sights & sounds of space shuttle history!

When he went into space for the first time, he remarked in a 2004 interview, "The sensation that I got initially was that from space you can't see discernable borders, and you begin to question why people don't like each other, because it looked like just one big neighborhood down there. The longer I was there, the greater my 'a citizen of' changed. The first couple of days, D.C. was where I concentrated all my views, and I was a citizen of Washington, D.C. … After two days, I was from America, looked at America as our home. … And after five or six days, the whole world became our home."

Sidney Gutierrez was the first Hispanic shuttle commander in April 1994 on *Endeavour* as part of STS-59, and John Herrington was the first tribal-registered Native American to fly and walk in space on STS-113 in 2002.

(Left) Ellison Onizuka, an Air Force test pilot, became the first Asian in space, launching on STS-51C. Onizuka only flew on one other mission, STS-51L, which ended in tragedy when *Challenger* exploded shortly after launch. (Opposite page, top right) Sharon McDougle checks the suit fit of Commander Fred Gregory before STS-44. (Opposite page, top left) Sid Gutierrez was the pilot on STS-40, the first Hispanic to pilot the shuttle, and later served as the first Hispanic shuttle commander on STS-59. (Opposite page, bottom) STS-113 mission specialist John Herrington, attired in his Extravehicular Mobility Unit (EMU) spacesuit, became the first tribal-registered Native American to walk in space in 2002.

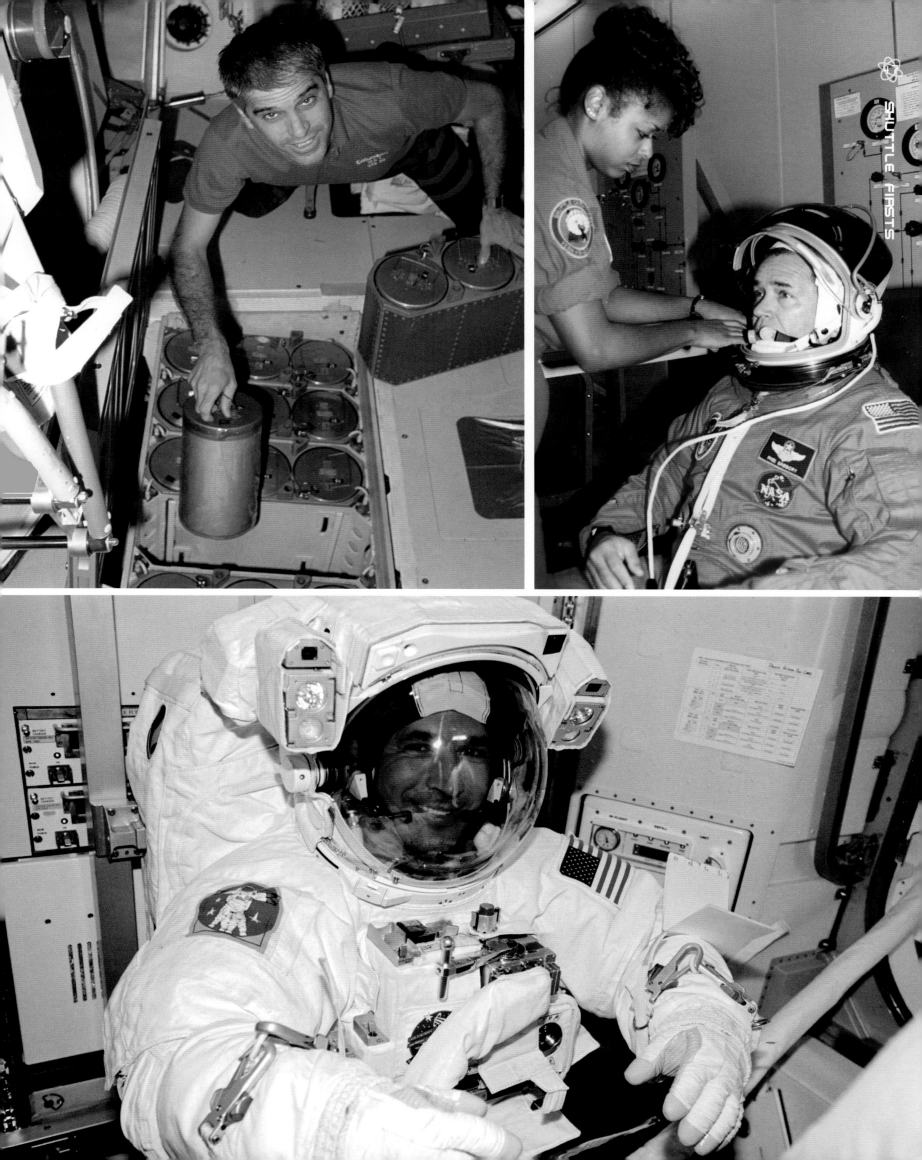

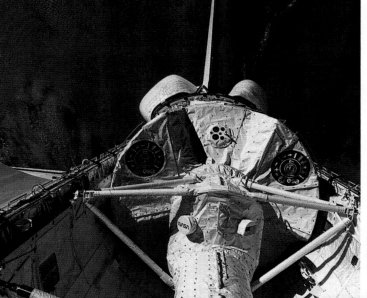

# Boundaries
# Extended

Unlike the Mercury, Gemini and Apollo missions that got the United States into space, and eventually to the moon, the space shuttle was meant to be an all-purpose craft that could help with science experiments, work with satellite deployment, retrieval and repair, and serve as a kind of space taxi to ferry astronauts to and from the Russian Space Station Mir and the International Space Station. The shuttle missions had plenty of other firsts from launch to landing.

Bruce McCandless was the first astronaut to perform an untethered spacewalk with the Manned Maneuvering Unit during February 1984 on *Challenger*. By doing so, he became the first human Earth-orbiting satellite.

The shuttle crews were very good at releasing satellites that they carried as cargo to space, but the crew of *Discovery* on STS-51A in November 1984 was the first to retrieve a satellite from Earth orbit as they plucked Palapa B-2 and Westar 6. Both were returned to Earth on the shuttle, refurbished and later relaunched.

In October 1985, the crew of the *Challenger* was part of the first dedicated German Spacelab mission, in which all activities were controlled from outside of the United States.

The crew on *Atlantis* released the first deep-space probe on a shuttle flight in May 1989 as Magellan was released and sent on its way to a rendezvous with Venus in 1990.

Another milestone was reached in February 1995 during STS-63 as the crew of *Discovery* made the first approach and fly around the Mir. Commander James Wetherbee maneuvered *Discovery* to within 36 feet of Mir during the mission in preparation for docking on a later flight.

**Hear Bruce McCandless as he tries out** the Manned Maneuvering Unit

Use your QR Code-Enabled device to see and hear the sights & sounds of space shuttle history!

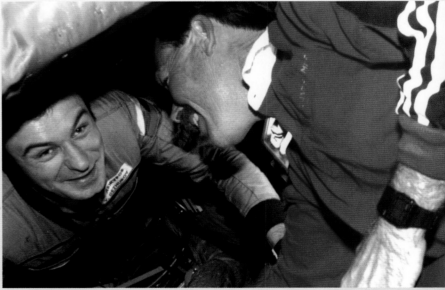

(Preceding page, bottom middle) STS-9 was the first NASA mission to be a cooperative venture among 11 European nations plus Canada, Japan and the United States. Spacelab featured over 70 different investigations in five research areas of disciplines: astronomy and solar physics; space plasma physics; atmospheric physics and Earth observations; life sciences; and materials science. (Left) Ulf Merbold from the European Space Agency flew aboard STS-9 to help monitor Spacelab. (Preceding page, top and bottom left) One of the new technologies that debuted during the shuttle era was the Manned Maneuvering Unit, a portable backpack that allowed an astronaut to do untethered spacewalks, essentially becoming a separate "spaceship." Astronaut Bruce McCandless was the first to use the MMU, flying out to 300 feet away from the shuttle on mission STS-41B. (Above) The shuttle was the first American spacecraft to dock with a Russian space station, when Atlantis docked with Mir on STS-71. Cosmonaut Vladimir Dezhurov was there to shake the hand of Commander Robert "Hoot" Gibson after opening the docking hatch.

### Hear STS-93 Commander James Wetherbee relay a message to Mir in both English and Russian

Use your QR Code-Enabled device to see and hear the sights & sounds of space shuttle history!

The shuttle and Mir finally docked when *Atlantis* did so during STS-71 in June 1995. The U.S. shuttle brought two relief cosmonauts to the space station, and brought back two cosmonauts and a U.S. astronaut, Norman Thagard, who were ending their stint on Mir.

The work with Mir during the nine missions where the two docked in space was a prelude to one of the shuttle fleet's primary jobs over its final 13 years: the assembly of the International Space Station in low Earth orbit.

A personal milestone was reached for one of NASA's original astronauts as U.S. Senator John Glenn, who was the first American man to orbit the Earth in *Friendship* 7 in 1962, went back into space 36 years later with STS-95 in October 1998 at age 77 to study the effects of space flight on the elderly.

In July 2005 on *Discovery*, the crew of STS-114 — the first after the re-entry accident that claimed the crew of *Columbia* — attained several firsts. Astronaut Stephen Robinson rode the robotic arm to inspect the orbiter's underside, the shuttle made a "rendezvous pitch maneuver" — a slow-motion back flip — to be photographed by ISS astronauts, and a demonstration of repair techniques on the shuttle's protective tiles was performed.

(Top left) Two cosmonauts, Anatoly Solovyev and Nikolai Budarin, rode up on STS-71 and stayed on as Mir's new crew while the previous crew of three returned to Earth on the shuttle. (Above) Another first for the shuttle program was the return to space of astronaut John Glenn, now a U.S. senator from Ohio, who made history for a second time as the program's oldest astronaut at age 77 when he lifted off with STS-95. (Opposite page, top left) Glenn participated in numerous experiments sponsored by the National Institute on Aging, which gathered information on bone and muscle loss, balance disorders and sleep disturbances. (Left) Jerry Ross tied a shuttle record with seven spaceflights when he lifted off with STS-110 in 2002. (Opposite page, bottom left and top right) Shuttle flights began doing checkouts of the spaceship's tiles on STS-114, the first mission after the loss of the *Columbia* on STS-107.

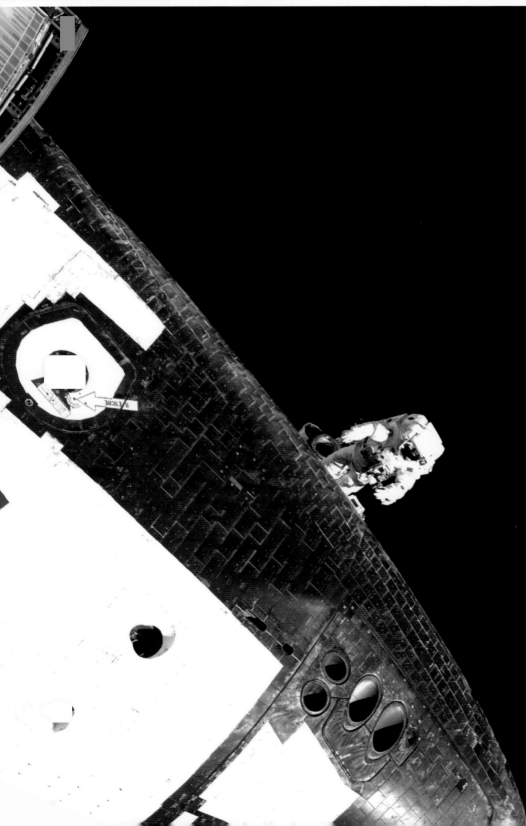

DANGER
DANGER
DANGER

# LIVING CONDITIONS

Astronauts orbiting our planet must live, sometimes for weeks, in an environment vastly different from the one they are accustomed to on Earth.

The far reaches of space expose travelers to extreme variations of heat and cold, and to deadly radiation. Floating hazards called micrometeoroids can also jeopardize vehicles with brunt-force collisions. The Earth's atmosphere protects us from the majority of these risks, but in space, humans are extremely vulnerable.

In this alien, weightless environment, astronauts must familiarize themselves with a completely new way of life where nothing can be taken for granted. Not only must they be kept safe from danger, but their most basic day-

to-day functions — breathing, waste elimination, drinking, eating, changing clothes and sleeping — must be provided for as well.

The International Space Station simulates the conditions found on Earth as closely as possible, as all ISS modules are designed for living, are well lit and the temperature is controlled. Because the work schedules of crewmembers are regimented and very challenging, it is important to the success of the mission that these men and women are kept comfortable.

NASA works hard to ensure that an astronaut's life on the ISS is as pleasant as possible. Despite this, life in space calls upon humans to make considerable adaptations. First-time astronauts must quickly become used to a weightless existence in cramped quarters, all the while carrying out the rigorous demands of the prescribed mission. These special conditions, common to all space stations, mean that even the most commonplace activities can call for added coordination, patience and scrupulous attention to detail.

(Preceding page, top) The shuttle's crew compartment measures in at 2,325 cubic feet, slightly smaller than a standard hotel room. (Preceding page, bottom left) Sending crews of anywhere from four to eight people up in such cramped quarters for two weeks at a time requires astronauts to treat their roommates with patience and understanding. Plus, in this orbiting "hotel room," both the gym (preceding page, bottom right) and the shower facilities (above) are public.

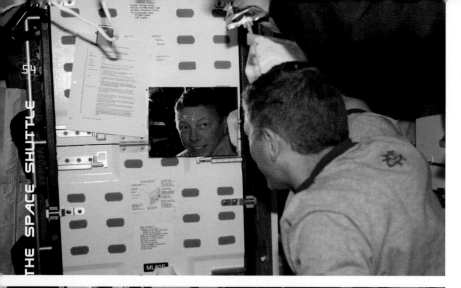

# Personal Hygiene in Zero Gravity

**D**uring the early days of space travel, the no-gravity environment — combined with extremely cramped quarters — made "space bathing" somewhat problematic.

Instead of showers, astronauts simply took sponge baths, using one wet cloth to wash and another to rinse. The larger International Space Station (ISS) was designed to include space for a full-body shower cabin. However, this is a shower that is found nowhere on Earth. Within the fully enclosed shower cylinder, each crewmember is limited to four liters of water to spray their bodies with a fine mist. They then apply special non-rinsing soap and shampoo. Since natural surface tension draws water and soap bubbles to the skin, very little water is actually needed. Within the shower cabin, crewmembers must wear a special breathing device to prevent breathing in the suspended water globules into their lungs. Due to zero gravity, it is necessary to smear the floating water droplets over the body, rather like applying suntan lotion. As soon as the droplets fasten to the skin, they form a clinging film that must then be wiped or scraped off. Lastly, astronauts simply vacuum up the water droplets suspended about the stall and towel themselves dry.

**See STS-92 pilot Pam Melroy wash her hair in orbit**

Use your QR Code-Enabled device to see and hear the sights & sounds of space shuttle history!

Collecting and getting rid of body wastes in zero gravity poses a major challenge to spacecraft designers. The functioning of a flush toilet on Earth depends upon gravity. On the

Personal hygiene in space is much different than keeping clean on Earth. Because floating water droplets can escape and end up damaging equipment, astronauts take sponge baths and use no-rinse shampoo. (Left, from top to bottom) Mike Fossum wets his hair for a "shower"; Heidemarie Stefanyshyn-Piper squeezes no-rinse shampoo into her hair; Lisa Nowak shows off her shampoo as she waits for her hair to dry; and Sandy Magnus uses a towel to speed up the drying process. (Opposite page, top) Robert Curbeam has less hair to wash, but male astronauts also have to deal with shaving (opposite page, inset). (Opposite page, bottom) Combing one's hair can be a crazy experience in space, as Tracy Caldwell demonstrates.

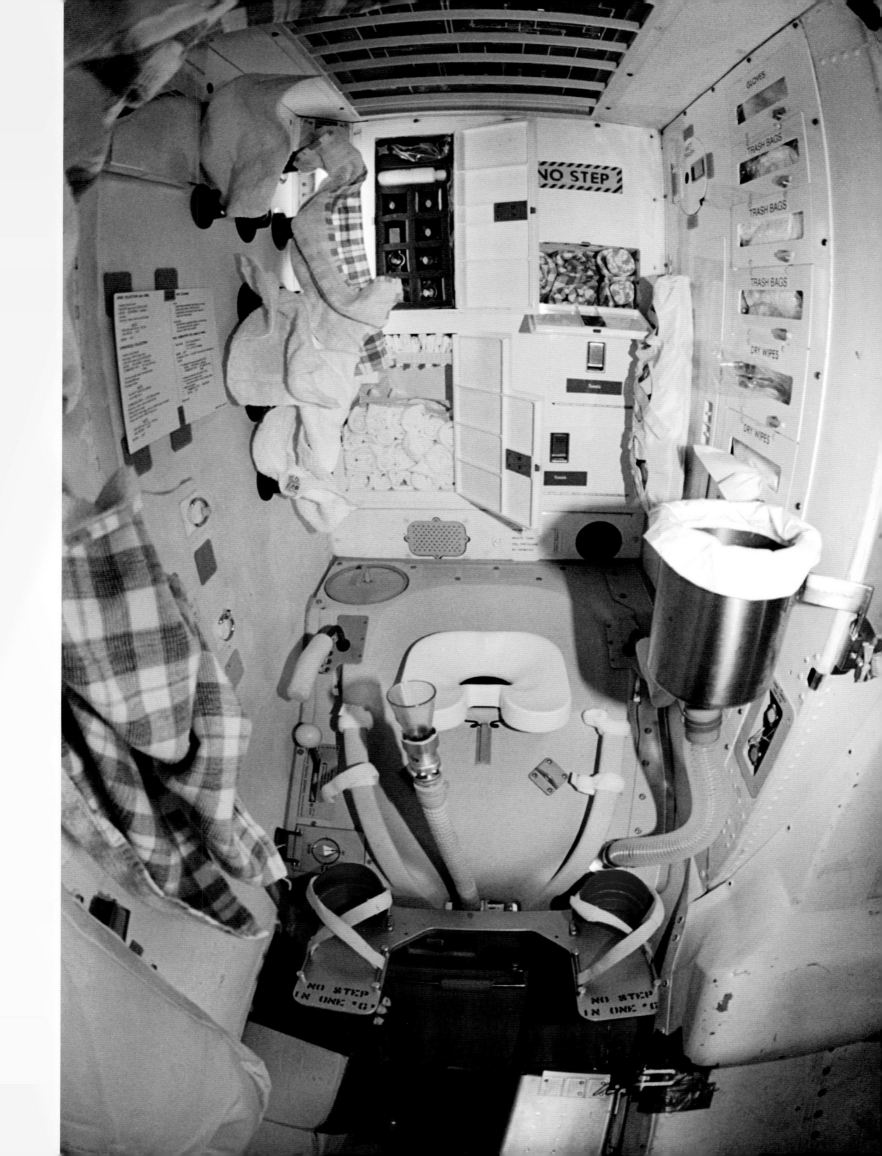

shuttle and the ISS, a no-gravity version had to be developed using suction, depending upon the flow of air rather than water. Astronauts strap themselves over an appliance that looks very much like a toilet seat. After usage, the flow of air moves the wastes into collection vessels located under the seat. When a task calls for working outside the spacecraft cabin, crewmembers simply wear special equipment that funnels body wastes into a pliable sack within their space suits.

On board the shuttle, the toilet procedure is usually referred to as the Waste Collection System (WCS). The usual airflow, combined with rotating fans, distribute solid waste into a cylindrical container for in-flight storage. It is then dried with vacuumed air to eliminate odor and pathogens. Liquid waste is vented to space. In the Zvezda and Tranquility modules of the ISS, two toilet facilities have been installed. These also operate on fan-driven systems. Unlike the WCS found on the shuttle, liquid waste is collected in 20-liter aluminum storage tanks for later removal.

During the July 2011 mission, ISS crewmembers described a suspicious odor emanating from one of the toilets of the orbiting station. NASA astronaut Ron Garan was assigned the task of repairing the zero-gravity toilet. Just a day earlier, he had successfully completed an important spacewalk vital to the success of the mission. The next day, he again exited the craft to repair and adjust the external waste-holding tanks. Garan's partner, Mike Fossum, later quipped, "That's the great thing about space flight. One day you're outside spacewalking, doing the most outrageous things that humans have ever done. The next day you're fixing toilets."

(Preceding page) The shuttle's Waste Collection System consists of a commode, urinal, fan separators, odor and bacteria filter, vacuum vent quick disconnect and waste collection system controls. When the commode is in use, it is pressurized, and transport air flow is provided by the fan separator. When the commode is not in use, it is depressurized for solid waste drying and deactivation. The urinal is essentially a funnel attached to a hose and provides the capability to collect and transport liquid waste to the wastewater tank. The fan separator provides transport air flow for the liquid. (Below) A door on the waste management compartment and two curtains attached to the inside of the compartment door provide privacy for crewmembers. One curtain is attached to the top of the door and interfaces with the edge of the interdeck access, and the other is attached to the door and interfaces with the galley, if installed.

**See crew habitability trainer Scott Weinstein** talk with astronaut Mike Massimino about toilet training in space

Use your QR Code-Enabled device to see and hear the sights & sounds of space shuttle history!

# Lighter-Than-Air
## Housekeeping

The men and women who operate the space shuttle and the International Space Station (ISS) are a highly trained and expert corps of astronauts, each with his or her individualized set of capabilities and responsibilities. However, when the crew is not carrying out tasks important to the success of the mission, they are also responsible for simple day-to-day housekeeping chores similar to those found in households on Earth. Living in a zero-gravity environment, though, sometimes means that the astronauts have to "relearn" how to perform even the most familiar of chores.

Efficient waste management plays an important role in the success of any space mission, and keeping the astronauts healthy during their time in orbit has always been one of NASA's top priorities. All trash consisting of damp organic matter, such as uneaten food, is quickly overcome by micro fungi and various molds. Shortly thereafter, gases and unpleasant odors can be detected if proper bagging and sealing methods are not employed. Without these methods, the buildup of gases in a garbage bag over several weeks could cause them to burst, emitting dangerous microbes into the cabin.

To prevent this possibility, crewmembers place items such as food waste into a container that allows these gases to slowly

**Hear STS-99 pilot Dominic Gorie talk about how much time** the crew has for housekeeping activities

Use your QR Code-Enabled device to see and hear the sights & sounds of space shuttle history!

(Top left) Pilot Kent Rominger demonstrates an age-old trash-compacting method on the middeck of the Earth-orbiting *Columbia* during STS-73. Following a meal, Rominger had collected the residue wrappers, and filled a plastic bag with them. Following his compacting maneuvers, Rominger went on to deposit the sack into a temporary trash-stowage area beneath the middeck. (Middle left) Pilot Steven Lindsey with a bag of refuse on *Discovery*'s middeck as the STS-95 crewmembers begin to settle in for a nine-day stay in Earth orbit. (Bottom left) Scott Parazynski, STS-95 mission specialist, with housekeeping chores on Flight Day 7 aboard *Discovery*. (Opposite page, top) *Endeavour* pilot Eric Boe uses a vacuum cleaner to remove dust particles from the air filter system on the shuttle's middeck. (Opposite page, bottom) Chris Hadfield of the Canadian Space Agency wipes clean one of the overhead windows on the aft flight deck of *Endeavour* as the seven-member STS-100 crew prepares for rendezvous with the ISS.

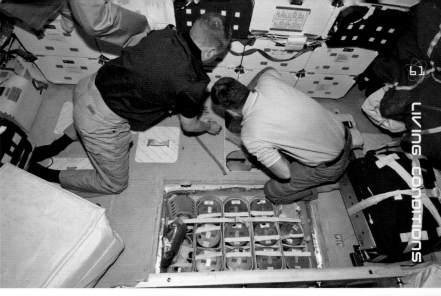

vent back into the cabin air-conditioning system where they can be quickly neutralized. Dry trash, such as containers and wrappings, are simply washed, dried and stored in large garbage units until the end of the mission. Upon returning to Earth, these units are removed and disposed of.

## Hear STS-113 mission specialist John Herrington talk about
cleaning up water spills in orbit

Use your QR Code-Enabled device to see and hear the sights & sounds of space shuttle history!

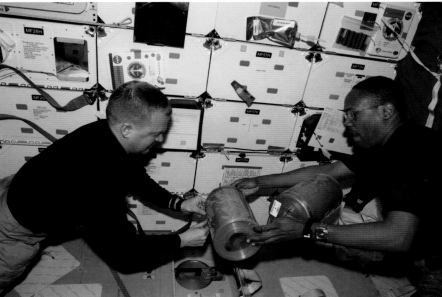

Keeping their living quarters clean and germ-free is an ongoing task. After a meal on Earth, clean up is easy — just wipe down the dining surface with a damp sponge or cloth. In space, things are not so simple. Any crumbs, food particles or stray liquids that escape their containers will free-float around the cabin. This is not only unsanitary, but might prove dangerous as well. Although the living compartments on the shuttle and the ISS are equipped with strong ventilation systems that constantly filter the air, it is necessary for the crewmembers to vacuum up particulate debris and dirt floating about the cabin or adhered to the cabin walls and fixtures. They use the same wet-and-dry vacuum cleaners to clean out the system air filters. After vacuuming, the cabin is typically wiped down with biocide, a liquid detergent that is sprayed onto all surfaces. For general housecleaning tasks, five wipes are used: wet, dry, fabric, detergent and disinfectant.

The possibility of future missions to the moon and even to Mars will mean increased lengths of time that astronauts live and work together in enclosed environments. The successful completion of these long-duration missions is directly related to the health of the crews within these challenging environments. To assure this, it is anticipated that future spacecraft will provide high-tech environmental health sensors designed to better monitor crew living areas and their atmosphere control systems.

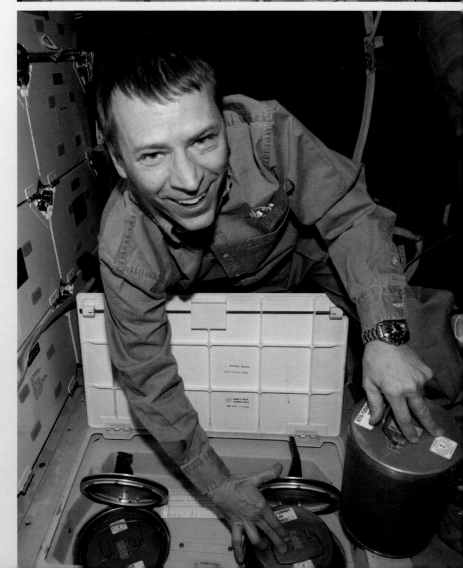

(Preceding page, top) Michael Fincke, STS-134 mission specialist, has some fun in the weightlessness of space while lifting heavy bags. (Preceding page, bottom) Japan Aerospace Exploration Agency astronaut Koichi Wakata, STS-127 mission specialist, is pictured near a lithium hydroxide canister floating freely on the middeck of *Endeavour*. (Top right) On *Discovery*'s middeck, STS-133 Commander Steve Lindsey, right, and Eric Boe work with lithium hydroxide canisters beneath the floor, performing the same housekeeping chore accomplished by many astronauts in the 30-year history of the program. (Middle right) Eric Boe and STS-133 mission specialist Alvin Drew move lithium hydroxide canisters. (Bottom right) STS-125 pilot Gregory C. Johnson works with lithium hydroxide canisters from beneath *Atlantis'* middeck.

# The
# Free-
# Floating
## Gym

**W**orking out and keeping fit is an important part of many American lives. Toned muscles, a healthy heart and reduced body fat are just a few of the benefits of maintaining a healthy lifestyle. For the crewmembers of the space shuttle and the International Space Station (ISS), sticking to a rigorous workout schedule is actually a NASA requirement.

The astronauts strive to maintain their fitness levels for all the same reasons they do on Earth, but there are many additional incentives that make working out in a zero-gravity environment absolutely necessary. Constant exercise is an important part of every astronaut's busy schedule, and during the course of their 16 waking hours, four hours are spent in scheduled exercise.

On Earth, a four-hour daily workout regimen might seem a bit a bit extreme. In zero gravity, however, crewmembers must keep very active or they would begin to suffer significant bone loss — effectively exhibiting all of the properties of osteoporosis. Whether we know it or not, our bodies are always working to counter the effects of gravity and keep us stable. Muscles and bones are constantly moving and flexing just so that we can remain standing or sitting upright. In a weightless environment, these bones are no longer needed to support us. This can cause skeletal unloading, a condition in which healthy bones will not only cease to create new bone cells, but process important minerals like phosphorus and calcium.

Research has shown that astronauts spending weightless months in space soon experience severely compromised bone mineral density (BMD). This is especially noticeable in the spine, neck and pelvis. On Earth, we lose about 3 percent of

### See video of an astronaut using the space shuttle's ergometer

Use your QR Code-Enabled device to see and hear the sights & sounds of space shuttle history!

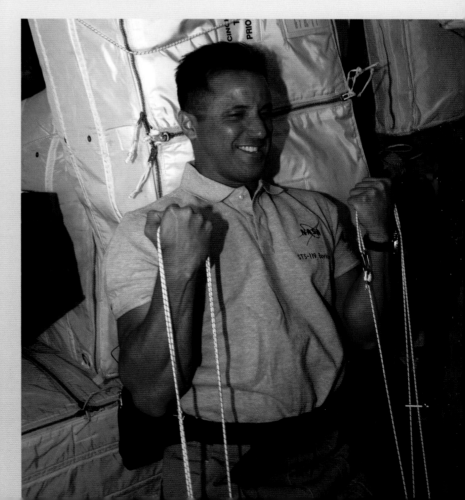

(Preceding page) Maintaining a rigorous workout schedule while in space is a NASA requirement. (Right) During their 16 waking hours per day, astronauts must spend four of those hours working out to prevent bone loss, which is magnified by being in space. (Top right) Mission specialist Steven Hawley runs on a treadmill on the middeck of *Columbia* on STS-93.

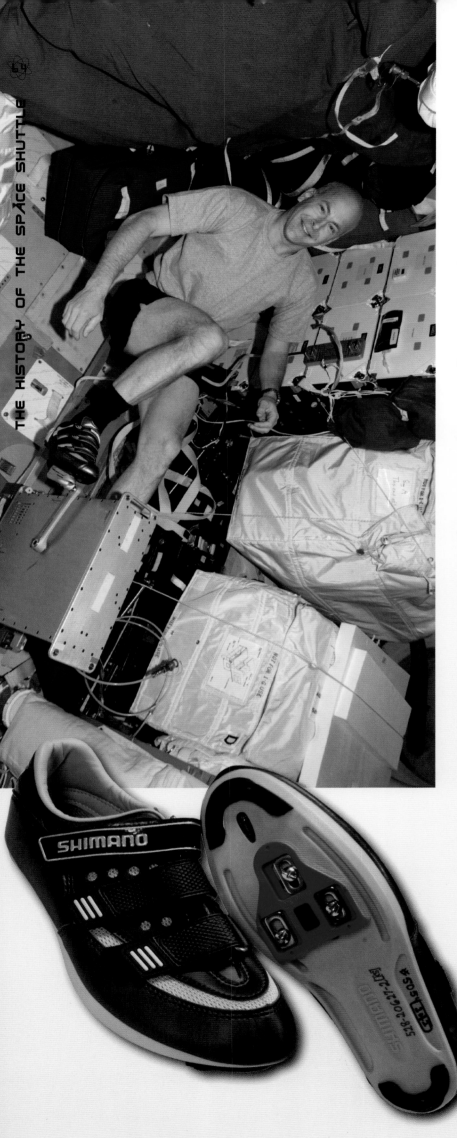

our cortical bone (the dense outer layer of the bone) during the course of about 10 years. In space, crewmembers can experience the same bone loss in less than a year. The chances of bone fracture upon return to Earth can be significant, and once bone loss is experienced, it cannot be recovered.

In space, astronauts must always combat the dangers of bone loss. One way is to utilize the ISS Treadmill Vibration Isolation System (TVIS). While similar to a treadmill found on Earth, there is one major difference — it is free-floating. The TVIS hovers about the living quarters in the same way the astronauts do. After a crewmember is harnessed into the TVIS and begins to "run," the treadmill moves with the runner.

## Hear STS-92 astronaut Peter Wisoff talk about exercising in space

Use your QR Code-Enabled device to see and hear the sights & sounds of space shuttle history!

The Cycle Ergometer with Vibration Isolation System (CEVIS) operates very much like a stationary bike. Bolted securely to the floor, crewmembers strap their feet to the pedals and wear a belt to hold them upright upon the seat. NASA has also developed the Restive Exercise Device (RED), which functions like a weight machine on Earth. Weights, however, are replaced with vacuum cylinders that can provide up to 600 pounds of pressure resistance.

While scheduled, intense workouts are a definite boon to the space traveler; astronauts spending more than a few months at a time in zero-gravity will always suffer at least a small amount of bone loss. While scientists continue to explore ways to combat this, currently they remain unsure if bone loss in space eventually tapers off or if it continues to grow.

(Top left) STS-122 pilot Alan Poindexter exercises on a bicycle ergometer on the middeck of *Atlantis* while docked with the International Space Station. (Left) After crewmembers put on the bicycle ergometer shoes, they can enjoy a good cardiovascular and leg-muscle workout. (Opposite page, top) STS-126 pilot Eric Boe exercises on a bicycle ergometer on the middeck of *Endeavour*. (Opposite page, bottom) NASA astronauts Sandy Magnus and Chris Ferguson practice setup of an exercise ergometer as the crew of STS-135 trains in the Crew Compartment Trainer (CCT) mock-up at NASA's Johnson Space Center on June 29, 2011.

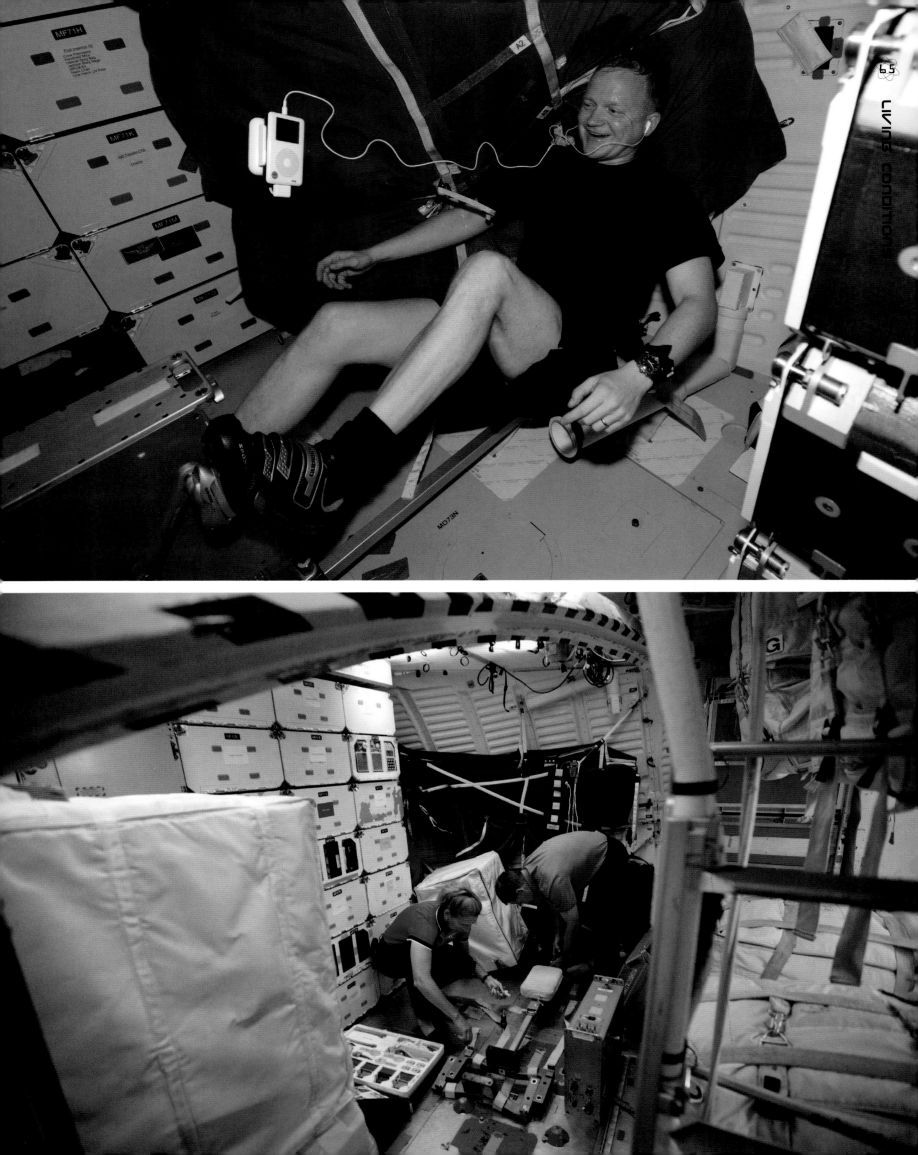

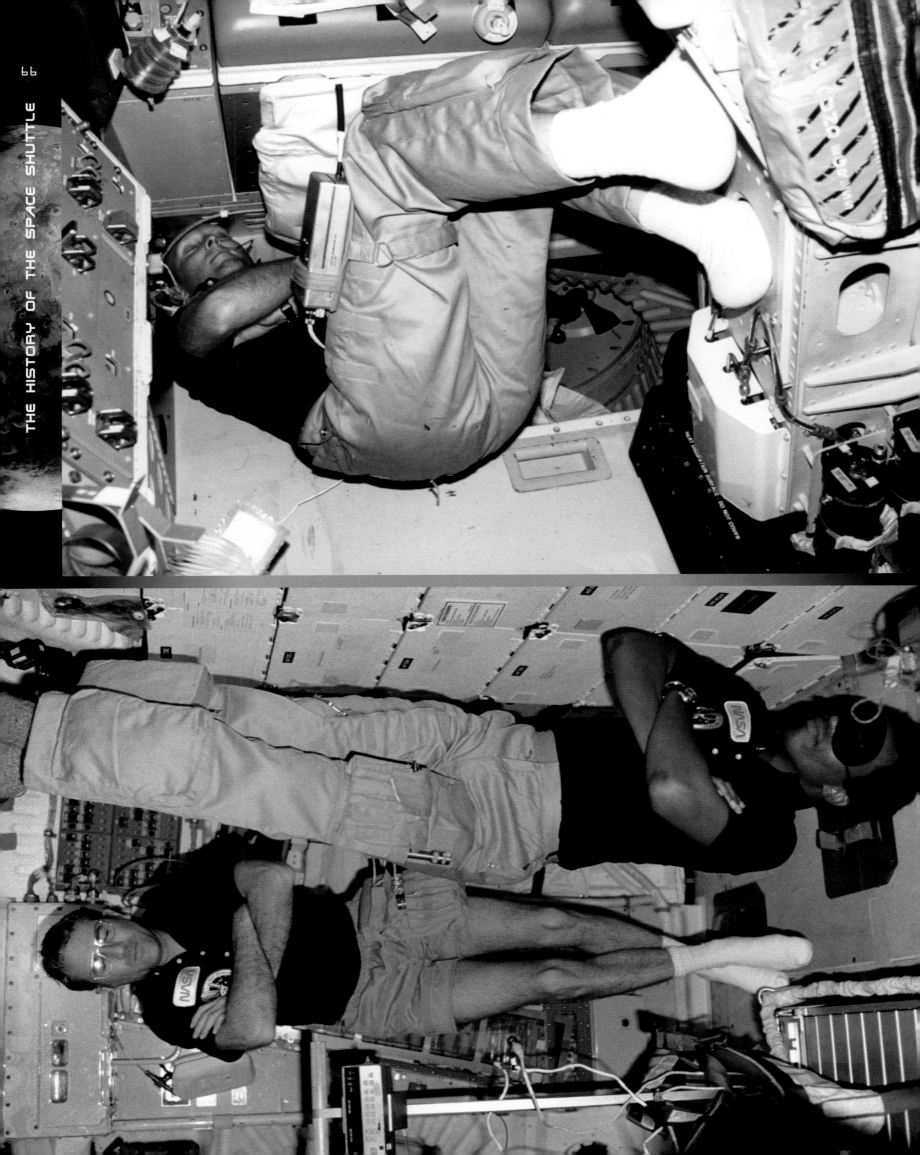

# Counting
# Zero-
# Gravity
## Sheep

**A**fter a long and regimented day of executing chores, spacewalking, exercising and research, most shuttle and International Space Station crewmembers are exhausted and more than ready for a good night's sleep. On Earth, the astronaut simply allows gravity to take over as his body goes limp atop a bed. But in the weightless environment in space, sleeping and sleep conditions must be looked at in a completely different light.

In zero gravity, astronauts do not need to "lie down" to sleep. Since they are weightless, the concept of lying down no longer exits. As long as they are tethered to something stationary, they can bed down any place or in any position they choose. Crewmembers on board the space shuttle will typically strap themselves into their seats. Some prefer to sleep in sleeping bags attached to the walls. Aboard the more spacious ISS, the astronauts can opt to sleep in vented compartments about the size of a broom closet. Inside this compartment, sleeping bags attach to the sides to prevent the sleeper from bumping into the walls and ceiling. A pillow, usually strapped to the head, helps to minimize bumps and bruises that might occur when the astronaut moves about in his sleep. NASA typically schedules in about eight hours of sleep per crewmember.

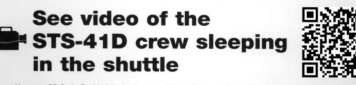

## See video of the
## STS-41D crew sleeping
## in the shuttle

Use your QR Code-Enabled device to see and hear the sights & sounds of space shuttle history!

(Preceding page, top) In the weightlessness of space, sleeping astronauts first tether themselves to a seat, a wall or an object. (Preceding page, bottom) STS-8 Commander Richard Truly and mission specialist Guion Bluford sleep in front of forward lockers and a portside wall. Truly sleeps with his head at the ceiling and his feet to the floor. Bluford, wearing a sleep mask, is oriented with the top of his head at the floor and his feet on the ceiling. (Top right) Astronauts can also opt to sleep in larger vented compartments on the International Space Station. (Right) Three STS-73 crewmembers are captured on camera at the end of their sleep shift on the middeck of *Columbia*. Pictured are, left to right, mission specialist Catherine Coleman, payload specialist Fred Leslie and mission specialist Michael Lopez-Alegria.

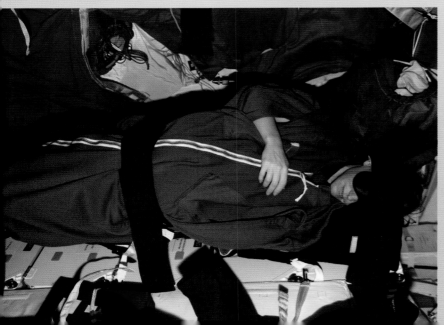

Adapting themselves to sleeping in zero gravity can take some time, but there are other problems that arise when trying to maintain regular sleep cycles in space. Our bodies, over thousands of years, have adapted themselves to a 24-hour cycle of waking and sleeping. Our innate response to this cycle is based largely upon the rising and setting of the sun. When orbiting the Earth, the occupants of the ISS are subjected to 16 sunsets each day — approximately one every 90 minutes. To counteract this disruption to their biological clock, NASA keeps the shuttle and ISS astronauts on a strict sleep schedule that adheres to Greenwich Mean Time.

## Hear STS-105 mission specialist Dan Barry talk about sleeping in space

Use your QR Code-Enabled device to see and hear the sights & sounds of space shuttle history!

Some sleep issues arise simply due to the emotional strain of being parted from loved ones. Some crewmembers experience motion sickness. On top of this, living in exceptionally tight quarters with several other astronauts can also cause sleep deprivation. For many crewmembers, sleep masks are a necessity to block out the rays of the sun and the lights of the living compartment. Earplugs are needed to drown out the day-to-day sounds of fellow astronauts and the whirring of the ship's ever-present air-filtration systems.

To counteract these problems, many astronauts rely upon sleep-inducing drugs. In the past, NASA reported that nearly half of the drugs prescribed to shuttle and ISS crewmembers are either sleep aids or hypnotic medications. Despite the use of these medications, some astronauts report that they actually get less than eight hours of sleep, with some claiming as little as six hours. On the other hand, other crewmembers report that their sleep patterns improve in space and that the sleep-related problems they commonly experience on Earth (such as sleep apnea) are greatly reduced or eradicated in zero gravity.

(Top left) The shuttle comes equipped with sleeping bags attached to the walls, and some astronauts choose to get their shut-eye in that manner. (Middle left) STS-120 Commander Pam Melroy, pilot George Zamka and mission specialist Paolo Nespoli sleep in their sleeping bags, which are secured on the middeck of *Discovery* while docked with the ISS. (Bottom left) STS-121 mission specialist Lisa Nowak sleeps in her sleeping bag, which is attached to the lockers on the middeck of *Discovery*. (Opposite page, top) STS-109 Commander Scott Altman sleeps on the flight deck of *Columbia*. (Opposite page, bottom) STS-125 pilot Gregory Johnson rests in his sleeping bag on the flight deck of *Atlantis*. (Opposite page, inset) U.S. Sen. John H. Glenn Jr., equipped with sleep-monitoring equipment, stands near his sleep station on the middeck of *Discovery* in 1998. The STS-95 payload specialist joined five astronauts and a Japanese payload specialist for nine days of a research mission in Earth's orbit.

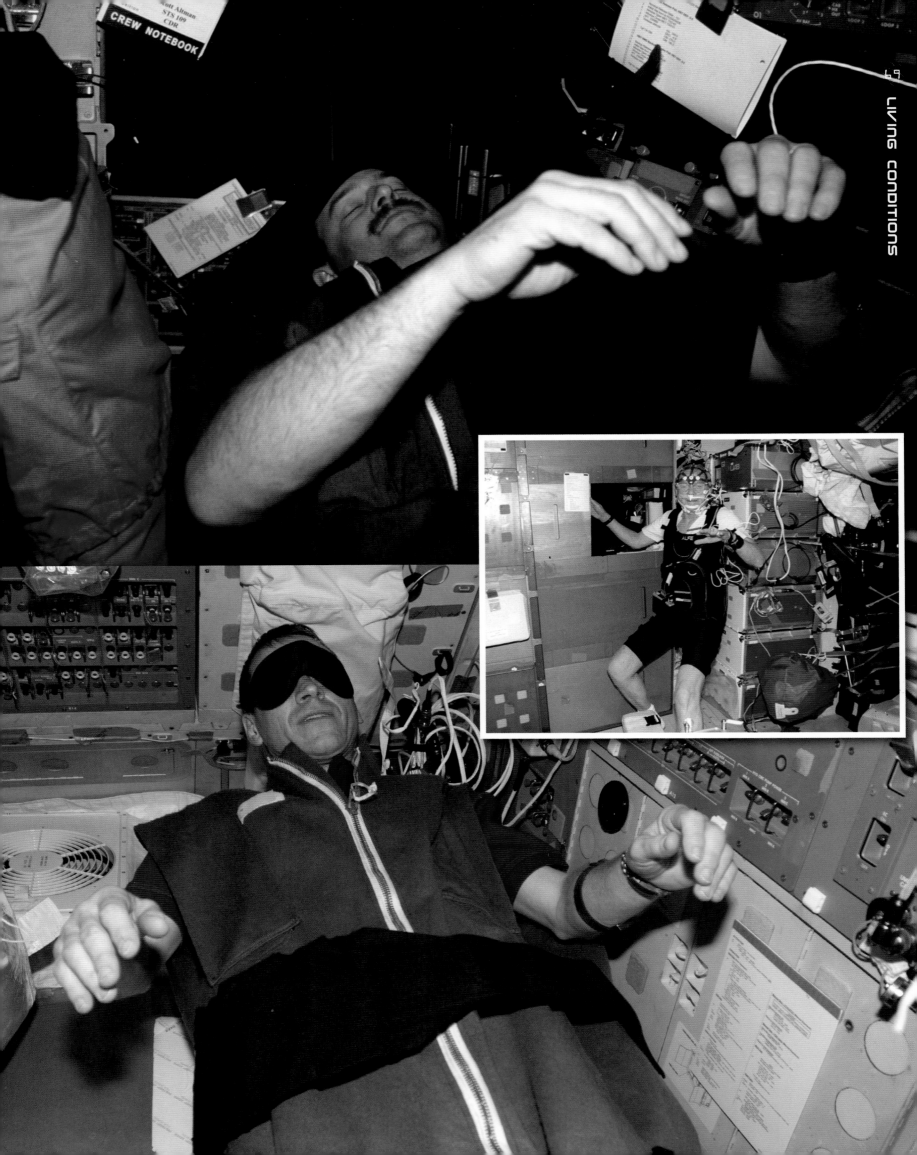

See astronaut Mike
Massimino and Vickie
Kloeris, manager of the
Space Food Systems Laboratory
at Johnson Space Center, discuss
the importance of food in space

Use your QR Code-Enabled device to see and hear the sights & sounds of space shuttle history!

# On THE MENU

Throughout history, one of the great problems to be overcome by explorers and travelers was developing methods for preserving and transporting foodstuffs for long-distance journeys.

Explorers like Magellan and Columbus preserved food using salt and brine. Hundreds of years later, canning and refrigeration addressed the problem. The advent of space travel in the early 1960s, however, demanded that new techniques be developed for keeping foods lightweight, compact, nutritious and appealing.

During the Mercury mission, astronaut John Glenn found himself forced to endure freeze-dried powders, jellylike substances packed in aluminum tubes and colorless, dry cubes.

Most crewmembers found that squeezing food from a tube to be disagreeable, and they all agreed that freeze-dried meals were difficult to rehydrate and caused flurries of floating crumbs that endangered some of their sensitive instruments.

During the later Gemini missions, NASA instituted several dining and menu improvements. Gone were the hated squeeze tubes, and prepared, diced foods were lightly coated in gelatin to help prevent crumbling. Freeze-dried foods were fully encased in plastic containers that made reconstituting them with water much more efficient. Gemini astronauts found that they had a variety of meals to choose from — some of the selections comprised of their favorite foods. The development of the Apollo program meant that food quality and variety would expand even further. Apollo crewmen were the first to enjoy hot water, which not only improved food flavor, but made rehydration much easier and less time consuming.

Today, NASA strives to ensure that space shuttle and ISS crewmembers experience eating and dining in much the same way as they do at home on Earth. Several ingenious steps have been taken to make certain that the zero-gravity environment does nothing to inhibit the experience of enjoying a tasty meal at the end of a hard day's work.

(Preceding page, top left) STS-133 mission specialist Nicole Stott prepares a snack on *Discovery*'s middeck — an apple and a tortilla, food items that do not create burdensome crumbs in the weightless environment of space. (Preceding page, top right) The development of the Apollo program led to improved food quality and variety, to the point where shuttle crewmembers just need to add a drop of water. (Preceding page, bottom left) STS-110 mission specialist Jerry Ross, along with a tray of food, floats on the middeck of *Atlantis*. (Preceding page, bottom right) STS-119 mission specialist John Phillips prepares to get ready to eat a meal near the galley on the middeck of *Discovery*. (Above) STS-128 pilot Kevin Ford holds a storage bag containing food items on the middeck of *Discovery* while docked with the International Space Station.

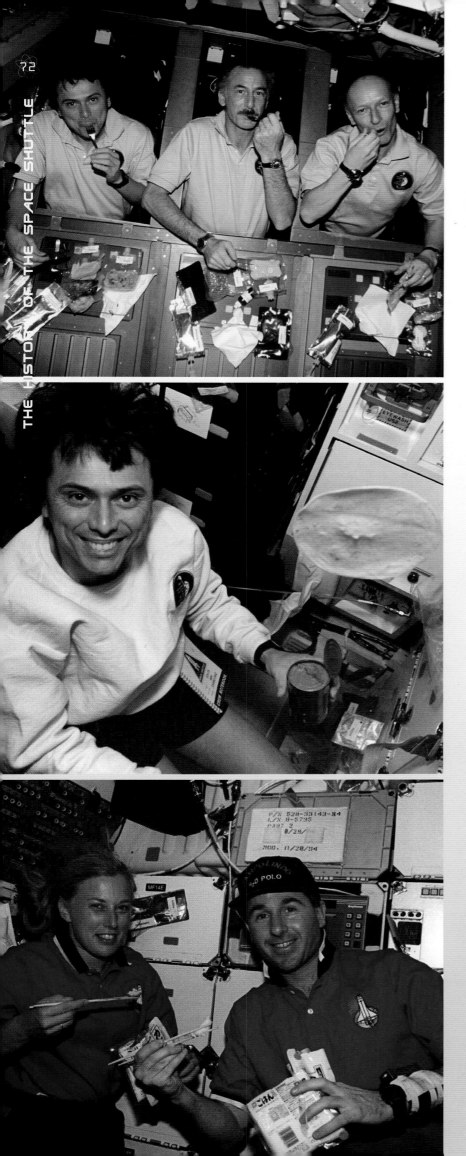

# Microgravity
# **Meals**

**M**any of us might be surprised that foods consumed by space shuttle astronauts are not strange, scientifically enhanced mixtures developed in laboratories, but simple, healthy fare readily available in grocery stores across the country. Still more surprising is the vast array of foods that crewmembers have to choose from. Selections are made from a NASA-designed menu plan, but alternative items can be substituted to fit the tastes of each crewmember after a trained dietician has made sure that dietary requirements are being met.

Located on the middeck, the space shuttle's galley is where all the meals are prepared. Simple and compact, it consists of several storage lockers, a water source for rehydrating food and an oven used for heating foods in their pre-packaged containers.

**Hear STS-111 Commander Ken Cockrell** talk about food in space

Use your QR Code-Enabled device to see and hear the sights & sounds of space shuttle history!

In the zero-gravity environment of space, sitting down to dinner at a table set with dishes, glassware and utensils is not an option. Most shuttle astronauts use a meal tray to keep all of the elements of their meal from becoming a "moveable feast." The food packages adhere to the tray, allowing crewmembers to enjoy several different kinds of food at the same time. Without them, they would be forced to eat just one food at a time. Knives, forks and spoons are used in space just as they are on Earth, but NASA has added one new utensil:

(Top left) Soon after reaching Earth orbit, the STS-75 blue shift team set up what they referred to as a "formal" meal on *Columbia*'s middeck. Enjoying the fine dining are, left to right, payload commander Franklin Chang-Diaz, and mission specialists Jeffrey Hoffman and Claude Nicollier. (Middle left) Franklin Chang-Diaz gets ready to prepare a tortilla on the middeck of *Columbia*. (Bottom left) STS-85 payload commander N. Jan Davis and mission specialist Stephen Robinson try using chopsticks while having a meal of Japanese rice on *Discovery*'s middeck. Robinson is wearing a special wristband indicating a Detailed Supplementary Objective (DSO) experiment. (Opposite page, top) Most of the food requires rehydration, which can take from two to five minutes. (Opposite page, bottom) STS-90 mission Commander Richard Searfoss sorts out food on the middeck of *Columbia*, using Velcro to attach the food packets to the trays mounted on the outside of middeck stowage lockers.

a small pair of scissors for cutting open the sealed plastic food containers. After the meal is over, the empty containers are rinsed and then stored in a trash compartment below the middeck level. The trays and eating utensils are simply wiped clean with a moist towelette.

 **See what STS-133 mission specialist Mike Barratt ate** for breakfast on day six of the two-week mission

Use your QR Code-Enabled device to see and hear the sights & sounds of space shuttle history!

Planning meals and provisioning the International Space Station (ISS) requires a different approach than the one used for the much smaller space shuttle, due primarily to the length of time crewmembers must live onboard. Every 90 days, the shuttle delivers a Multi-Purpose Logistics Module (MPLM) to the ISS filled with water and food supplies. Water is an especially valuable commodity. On the shuttle, the craft's electrical system produces byproduct water. Since the ISS is run primarily by solar panels, this is not an option. Water is recycled from condensate caused by the humidity in the air and by processing the urine of the crewmembers.

Should an emergency occur on board the ISS, NASA has developed the Safe Haven food system that will provide astronauts with 22 days of food and water. Designed to be completely independent of the day-to-day provisioning of the crew, the Safe Haven food system is light, takes up a minimal amount of space and can provide crewmembers with 2,000 calories a day.

Most of the ISS food containers require rehydration, a process that takes anywhere from two to five minutes. Next, the container is heated and then cut open with scissors. Contents can be eaten with a spoon since the moisture helps the food adhere to the spoon. As on the shuttle, meals provide a time to relax and chat with other crewmembers, watch satellite television or chat with family members back home via satellite phone.

(Preceding page, top left) STS-110 mission specialist Ellen Ochoa prepares a meal on the middeck of *Atlantis*. (Preceding page, top right) STS-119 pilot Tony Antonelli watches a spoonful of food float freely while he eats on the flight deck of *Discovery*. (Preceding page, bottom) STS-110 mission specialist Ellen Ochoa and Commander Michael Bloomfield eat a meal on the middeck of *Atlantis*. (Top right) European Space Agency astronaut Paolo Nespoli, a STS-120 mission specialist, poses for a photo while his spoon and food package float freely nearby on the middeck of *Discovery*. (Middle right) STS-128 mission specialists Jose Hernandez, left, and John "Danny" Olivas prepare tortillas near the galley on the middeck of *Discovery*. (Bottom right) STS-134 mission specialist Roberto Vittori from the European Space Agency enjoys a sandwich onboard *Endeavour*.

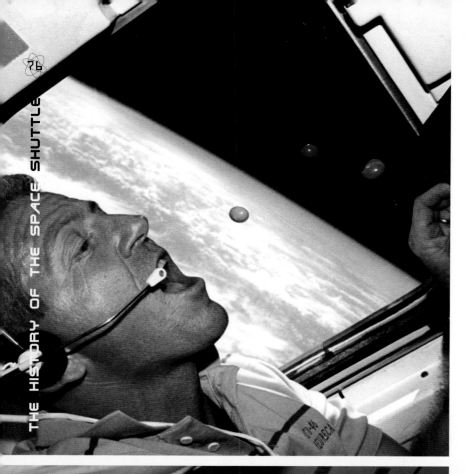

# Snack Attack
## at 1,300,000 Feet

In addition to the planned three meals per day NASA includes on the space shuttle, it also provides the crewmembers with an on-board pantry that offers a two-day contingency food supply, extra beverages and snacks. The pantry holds a hodge-podge of snack items that are quite popular, such as butter cookies, granola bars, peanut butter, trail mix, beef jerky, almonds and cashews. All of these are easily eaten in bite-sized portions without leaving crumbs — a big plus in orbit.

"In general, chips don't work well," said Vickie Kloeris, manager of the Space Food Systems Laboratory at the Johnson Space Center. "Some crewmembers have chosen to take the Pringles as part of their fresh/special request food since they are in a can and can withstand more than a bag of chips. However, they still make a lot of crumbs and the crewmember has to be willing to put up with that and likely will be forced by fellow crewmembers to clean up the mess they leave behind."

Pringles aren't the only commercially available product to have gone up on the shuttle. On various missions, astronauts have requested and flown M&M's, Goldfish crackers, Little Debbie Brownies, Lifesavers and Trident sugar-free gum. President Ronald Reagan, famous for his love of jelly beans,

**See STS-41C mission specialist James van Hoften take a bite** out of a spinning banana

Use your QR Code-Enabled device to see and hear the sights & sounds of space shuttle history!

(Top left) Not only do they melt in your mouth, but they also float in space! M&M's are just one commercial food product that has made the journey on a shuttle, along with Pringles potato chips, Goldfish crackers and, of course, Little Debbies. (Bottom left) STS-114 Commander Eileen Collins watches a container of vanilla pudding float on the mid-deck of *Discovery*. (Opposite page, top) President Ronald Reagan sent some jelly beans along on STS-7. The label on the plastic bag says, "Compliments of the White House." Pictured in front are Dr. Sally Ride and Dr. Norman Thagard, and in back are Commander Robert Crippen, pilot Frederick Hauck and mission specialist John Fabian. (Opposite page, bottom) STS-101 pilot Scott Horowitz experiments with peanut butter in space.

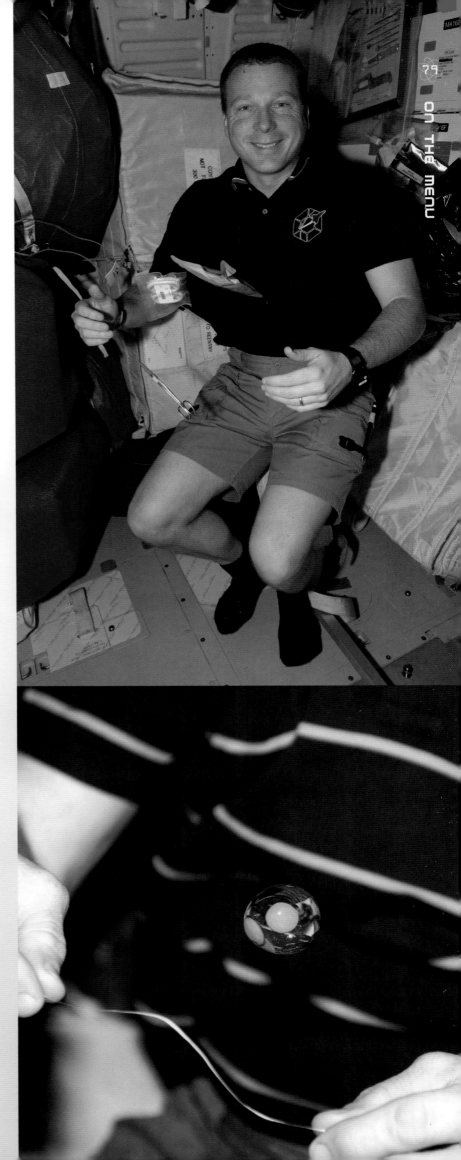

sent a package of Jelly Bellies along on STS-7 as a special treat for the crew.

Shuttle astronauts tend to snack, not because they're bored or need a quick pick me up, but because snacks are quicker than eating full meals.

"Shuttle crewmembers tend to skip meals a lot, they'll tend to snack rather than really eat a meal," said Kloeris. "They tend to especially skip lunches because they are so busy during the day. Their mission is only about two weeks long and they feel compelled to do as much as they can."

In that stressful work environment, one far from the creature comforts of home, indulging in a brief "snack break" can also provide crewmembers a chance for a bit of light-hearted fun. Astronauts have been known to do Pac-Man routines with M&M's and channel their inner shark by gobbling up "swimming" Goldfish crackers.

---

 **See STS-123 mission specialist Bob Behnken play** with some jelly beans in orbit

Use your QR Code-Enabled device to see and hear the sights & sounds of space shuttle history!

---

STS-41C mission specialist James van Hoften, known as "Ox" to his crewmates, is the star of one of the most famous examples of on orbit food fun.

"I love bananas, so I told these guys I wanted to have one every day," said van Hoften. "I was up there, and I peeled this banana, and all the little peels came down, and I said, 'Gee, this looks like a satellite, you know, if you spin it and stabilize it.' We were laughing at that, and then I went over and I took a bite out of it.

"About that time [Commander Bob] Crippen came down, and he said, 'Whatever you do, don't put that on film.' We looked at each other and go, 'OK, sure. We won't, boss. No problem.' As soon as he left, we spun up another banana, and I floated over and took a bite out of it."

---

(Preceding page, top left) STS-118 mission specialist Dave Williams, representing the Canadian Space Agency, prepares to eat a snack on the middeck of *Endeavour* while docked with the International Space Station. (Preceding page, top right) STS-122 mission specialist Daniel Tani holds a bag of candy while watching several pieces floating freely on the middeck of *Atlantis*. (Preceding page, bottom) STS-122 mission specialist Leland Melvin witnesses microgravity in action on the aft flight deck of *Atlantis* while docked with the ISS. A package of peaches, a scissors and a spoon float in front of Melvin. (Top right) Terry Virts, STS-130 pilot, takes a snack break on the middeck of *Endeavour*. (Bottom right) STS-123 Commander Dominic Gorie holds a piece of string near a water bubble with candy trapped inside as it floats freely on the middeck of *Endeavour*.

# FUN in SPACE

Flying a space shuttle mission is highly complicated and stressful endeavor, and by the time the shuttle actually lifts off, the astronauts have already been training hard for months. But astronauts like to play, too, and as shuttle flights grew longer, crews found plenty of creative ways to inject a little fun into their time in space.

Every morning when it's time for the astronauts to get up, Mission Control plays wake-up music, which has included specialized comedy bits by Robin Williams and Houston DJs, and parodies of popular rock songs with lyrics relating to that day's mission objectives.

(Preceding page, top) Rick Hieb, a mission specialist aboard STS-49, looks into the aft flight deck of *Endeavour* during his spacewalk. STS-49 marked the first shuttle mission to feature four EVAs. Hieb, along with fellow astronauts Pierre Thuot and Thomas Akers, helped to recover INTELSAT VI, a communications satellite whose orbit had become unstable. (Preceding page, bottom left) Big Head Todd & The Monsters play live in Mission Control for the STS-133 crew's wakeup song. It was originally written as a tribute to the space program and workforce. The live performance was the first time a shuttle crew has been awakened live from Mission Control in Houston. (Preceding page, bottom right) STS-112 mission specialist Sandra Magnus' idea of fun was working out on a bicycle ergometer on the middeck of *Atlantis*. (Above) Disney's space ranger Buzz Lightyear returned from space Sept. 11, 2009, aboard *Discovery*'s STS-128 mission after 15 months on the International Space Station. Buzz is part of NASA's series of educational online outreach programs.

As the crew goes about completing those mission objectives throughout the day, there's plenty of radio chatter between the shuttle and Mission Control, and crews don't hesitate to tease each other or play a practical joke now and then. Sometimes even the crew's work is play, such as experimenting with different children's toys to see how they respond in microgravity. Buzz Lightyear, the space ranger from the movie "Toy Story" has even flown aboard the shuttle.

Crews make silly signs and creative poses for their in-orbit group photo. Astronauts fly interesting items, such as a Major League Baseball home plate, in their personal preference kits while NASA flies items like movie props from the "Star Wars" series and the Olympic torch in the official flight kit.

And for some space flyers, just staring out the window as the Earth spins beneath them is enough to keep them enthralled.

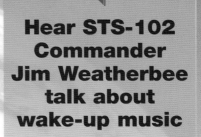

## Hear STS-102 Commander Jim Weatherbee talk about wake-up music

Use your QR Code-Enabled device to see and hear the sights & sounds of space shuttle history!

# Say, "Tracking and Data Relay Satellite!"

The early shuttle missions were busy affairs, packed with checkout tests for the vehicle and other mission objectives, and the astronauts didn't have much time for fun or practical jokes. However, taking photographs was already a large part of the astronauts' mission and it didn't take too much time to snap a fun or interesting picture every now and then.

Astronaut Joe Allen, a self-admitted pusher of the rules, may have started a trend of funny photos on shuttle missions when he snuck an old-fashioned shutter-release mechanism onto STS-5, just to capture a photo of all four of the first four-person crew in orbit.

"In the flight photos that came back, there were numbers of photos of us, the four crewmen," said Allen. "But this was with a camera that had no delayed shutter release! Not one NASA person said a word to me about it, but you knew that the people in the photo shops wondered how in the world those photographs were taken."

Not all astronauts were as willing as Allen to sneak unauthorized equipment on board the shuttle, but some were willing to carry on another photo tradition that Allen started on STS-5.

(Top left) On the first four-person shuttle mission, STS-5 mission specialist Joseph Allen snuck an old-fashioned shutter-release camera onboard to capture a team photo. Also pictured with Allen are Commander Vance Brand, pilot Robert Overmyer and mission specialist William Lenoir. The mission objectives included the deployment of two communication satellites, thus the "Ace Moving Co." sign. (Middle left) The five-member 51-A crew celebrates a successful mission. From left to right in front are David Walker, Anna Lee Fisher and Joseph Allen, and in back are Dale Gardner and Frederick "Rick" Hauck. The reference to "the Eagle" on the sign has to do with the *Discovery* crew's mascot. (Bottom left) The STS-37 crew also had signs to signify their goals and accomplishments on *Atlantis*. Commander Steven Nagel and pilot Kenneth Cameron are in back, and in front are mission specialists Jerry Ross, Linda Godwin and Jerome Apt. Ross holds a sign that reads, "Ace Railroad, STS-37" to pay tribute to the thorough evaluations of crew and equipment translation aid (CETA) carts that he and Apt conducted during the mission's second extravehicular activity (EVA). Godwin holds the STS-37 crew insignia, and Apt shows off a giant-sized ace of spades with an overlay that reads, "Ace Observatory Co.," in reference to the deployment of the Gamma Ray Observatory. (Opposite page, top left) In *Discovery*'s airlock, pilot William Readdy holds up a STS-51 slogan: "Ace HST Tool Testers." Readdy is flanked by Carl Walz, left, and James Newman. (Opposite page, top right) Dale Gardner and Joseph Allen display a "For Sale" sign while retrieving two satellites from orbit after their Payload Assist Modules (PAM) failed to fire. (Opposite page, bottom) Gardner provides a closer look at the "For Sale" sign while Allen is reflecting in Gardner's helmet visor.

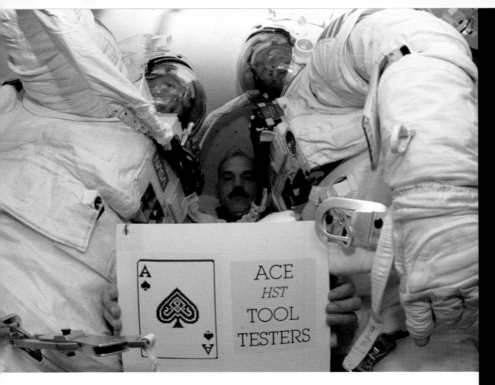

"Joe Allen coined the term 'Ace Moving Company,' because we moved stuff to space," said Commander Vance Brand. "We put up a sign, 'Ace Moving Company,' that Joe had made out ahead of time."

The crews of STS-37, STS-51 and STS-51A all came up with their own variations of the "Ace Moving Company" theme, and Allen took the joke to the next level on 51A, when he and fellow spacewalker Dale Gardner were sent out to retrieve the malfunctioning Palapa B-2 and Westar 6 satellites. Because the satellites were to be returned to Earth,

**Hear the STS-5 crew start** the "Ace Moving Company" theme by declaring "We deliver!"

Use your QR Code-Enabled device to see and hear the sights & sounds of space shuttle history!

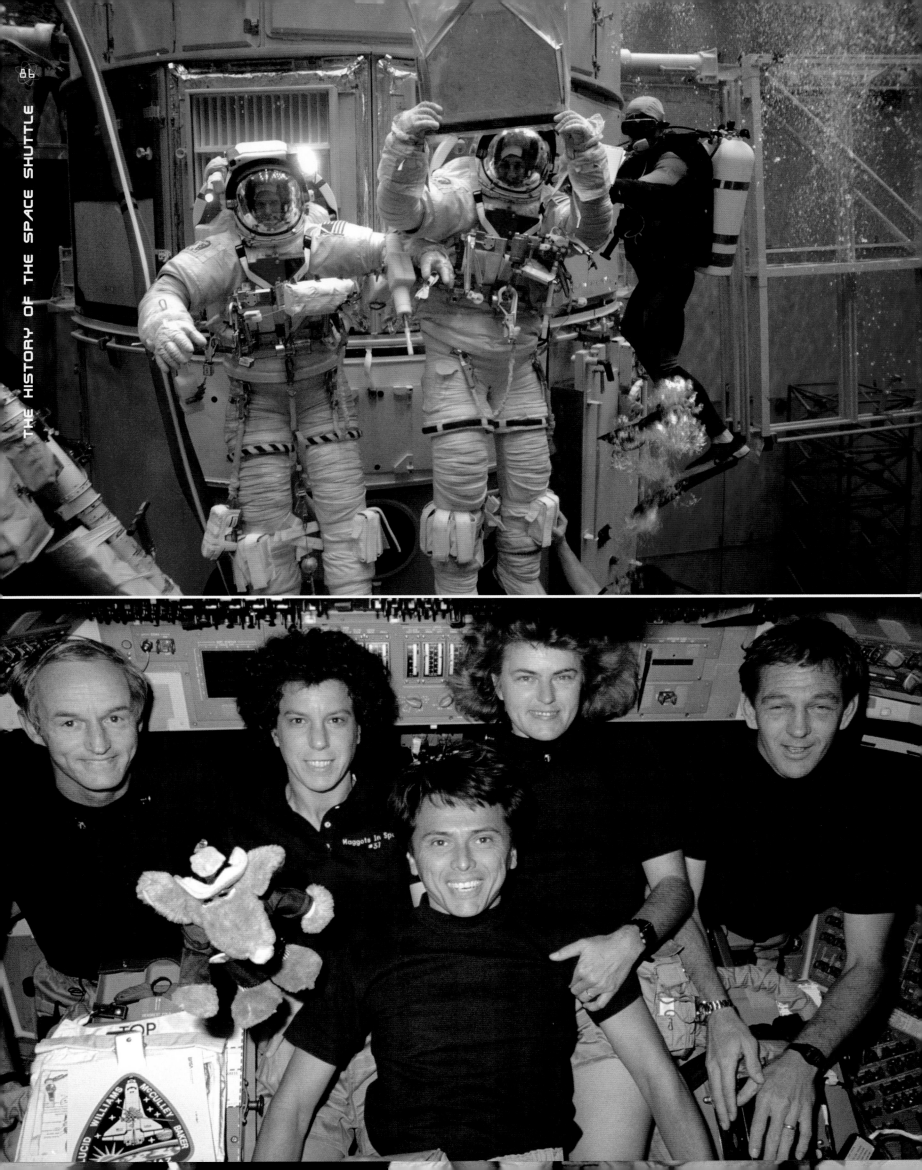

# Souvenirs From Space

**F**lying into space and back is the trip of a lifetime, and though astronauts are trained to be the ultimate professionals, they like bringing back a few souvenirs from their trip. Of course, there are no tacky tourist shops at 200 miles above the Earth, so souvenirs have to be carefully packed and taken along.

Each crewmember is issued a Personal Preference Kit, a 5-inch by 8-inch by 2-inch bag in which they can carry up to 20 items that must weigh less than 1.5 pounds. Crewmembers must submit a list of items for approval at least 60 days before launch and must arrive at the Johnson Space Center at least 45 days prior to the flight in order for them to be listed on the cargo manifest, packaged, weighed and stowed aboard the shuttle.

Astronauts generally choose to take things that have sentimental value to them already: Bob Behnken took along wedding rings for his upcoming marriage, Mike Fossum took some of his sons' Boy Scout badges and pins, and Eileen Collins flew a scarf belonging to fellow aviation pioneer Amelia Earhart. Jerseys and sports memorabilia are popular choices, too, such as when New York Mets fan Mike Massimino took along a home plate from Shea Stadium and Garret Reisman flew a vial of dirt from Yankee Stadium.

Some crews even flew items to poke some fun at the folks back in Mission Control. The STS-34 crew, plagued by problems with the communications equipment during training, adopted a stuffed toy rat as their unofficial crew mascot, a Mr. Ratty Comm. Ratty became such a part of the crew that he made the trip into orbit and even earned a spot in the crew's on-orbit photo.

**Watch STS-54 mission specialist Greg Harbaugh dunk** a miniature basketball onboard the shuttle

Use your QR Code-Enabled device to see and hear the sights & sounds of space shuttle history!

(Preceding page, top) New York Mets fan Mike Massimino, right, and Mike Good display home plate from Shea Stadium during a break from their training in the Johnson Space Center Neutral Buoyancy Lab. The astronauts took the plate with them when boarding *Atlantis* for STS-125. (Preceding page, bottom) STS-34 crewmembers pose for an onboard photo on *Atlantis*. From left to right are Commander Donald Williams, Ellen Baker, Franklin Chang-Diaz, Shannon Lucid and Michael McCulley. Also pictured is a stuffed toy rat, their unofficial mascot. (Top right) Mario Runco Jr. wears the jersey of the Houston Aeros, a minor-league hockey club, on STS-77. (Middle right) Pat Forrester gives a 400-year-old shipping tag from Jamestown, Virginia, to Elizabeth Kostelny from the Association for the Preservation of Virginia Antiquities after it made the journey to space and back. (Bottom right) STS-101 crewmembers, from left, James Voss, Susan Helms, Yury Usachev, James Halsell Jr., Scott Horowitz, Jeffrey Williams and Mary Ellen Weber hold aloft a replica of the Sydney 2000 Olympic Torch. The international crew flew a similar torch onboard *Atlantis* during its mission to the ISS to pay tribute to the international nature of the Olympic Games.

NASA gets in on the act, too, with the Official Flight Kit, 2 cubic feet of space reserved for carrying official mementos for the agency and other organizations such as aerospace contractors, NASA's external payload customers, other federal agencies, state and local governments, the academic community and independent business entities.

Many of the items NASA chooses to included in the OFK have real and/or symbolic links to space exploration, such as a 400-year-old shipping tag from Jamestown, one of America's earliest settlements, or a lightsaber prop from the "Star Wars" movie franchise. The "Star Wars" movies were a big inspiration to many of the astronauts, scientists and engineers in the space program, and those men and women have worked hard to make some of the fantasy become reality.

## See STS-125 mission specialist Mike Massimino talk

about taking Shea Stadium's home plate into space

Use your QR Code-Enabled device to see and hear the sights & sounds of space shuttle history!

"There's a kind of a fine line between science fiction and reality as far as what we do, and it's only just time really because a lot of what we're doing right now was science fiction when I was growing up," said astronaut Jim Reilly.

Kids today may not be as familiar with Luke Skywalker, but Disney's toy spaceman, Buzz Lightyear, flew aboard the shuttle and lived on the International Space Station — hopefully helping a new generation of students to get excited about spaceflight.

(Preceding page, top) STS-101 crewmember Yury Usachev from the Russian Aviation and Space Agency holds a replica of the Sydney 2000 Olympic Torch on the middeck of *Atlantis*. (Preceding page, middle) With an official National Football League football free-floating in front of them, the STS-27 crew poses for their inflight portrait. Pictured are, left to right, Commander Robert Gibson, Richard Mullane, Jerry Ross, William Shepherd and Guy Gardner. The football was later presented to the NFL at halftime of the Super Bowl in Miami. (Preceding page, bottom left) Baseball caps from the two 1995 World Series representatives float near a cabin window on *Columbia*, with Earth in the background. The American League champion Cleveland Indians and their National League counterpart Atlanta Braves were engaged in a scheduled best-of-seven World Series throughout the first portion of the 16-day mission in space. (Preceding page, bottom right) The STS-73 crew also brought along official National League and American League baseballs for the 1995 World Series. (Top right) Astronaut Mike Good, right, presents a Cleveland Cavaliers jersey to Cavs GM Chris Grant after taking it along for a ride on STS-132. (Middle right) James Voss, left, and James Kelly share a friendly moment onboard the International Space Station's U.S. laboratory Destiny, in spite of the long-standing rivalry between their respective alma maters — Auburn University and the University of Alabama. (Bottom right) Stephanie Wilson and Michael Fossum, STS-121 mission specialists, pose for a photo on the aft flight deck of *Discovery*. Fossum, a graduate of Texas A&M University, flashes the traditional "Gig 'Em, Aggies!" sign and wears an A&M cap, while University of Texas alum Wilson gives the "Hook 'Em, Horns!" sign and wears the Longhorns cap.

RIGHT LIGHT MODULE
COVER ASSEMBLY
P/N SED33111721-712

# BLAST
## OFF TO WORK

In 1981, NASA introduced its new manned space vehicle — a delta-winged orbiter, two solid rocket boosters and a large orange external fuel tank — as the Space Transportation System, but the media and the public quickly dubbed it the "space shuttle." Unlike the Apollo command and lunar modules, with specific vehicles designed for specific tasks and journeys, the shuttle was designed to be exactly what its name implies: a reusable vehicle available to ferry crewmembers, supplies, satellites, construction materials and scientific experiments into low Earth orbit.

Since its first flight, the shuttle has carried aloft more than 3 million pounds of cargo, and deployed 180 objects into space, everything from the huge 25-ton Chandra X-Ray Observatory on STS-93 to the PicoSat, a 5- by 10-inch satellite weighing 8 pounds, deployed by the final mission of the shuttle era. Perhaps the most famous, or infamous, piece of cargo the shuttle has carried is the Hubble Space Telescope, which, after a mission to correct a faulty lens, has produced some of the most visually stunning and beautiful images of the universe ever taken.

Hubble wasn't the only satellite that needed a bit of help, and the shuttle's airlock offered easy access for suited spacewalkers to capture, repair and upgrade hardware.

Even when there were no spacewalks scheduled for a mission, there was still plenty of work to be the done, as the shuttle hosted more than 2,000 experiments in the fields of Earth, astronomy, biological and materials sciences.

---

(Left) STS-123 mission specialist Rick Linnehan uses a digital camera to expose a photo of his helmet visor during construction and maintenance on the International Space Station. Also visible in the reflections in the visor are various components of the ISS, the docked *Endeavour* and a portion of Earth. There's always plenty of work to do on a shuttle mission, as crewmembers hosted more than 2,000 experiments in addition to hauling cargo and construction materials.

## Hear STS-98 pilot Mark Polansky talk about working in space

Use your QR Code-Enabled device to see and hear the sights & sounds of space shuttle history!

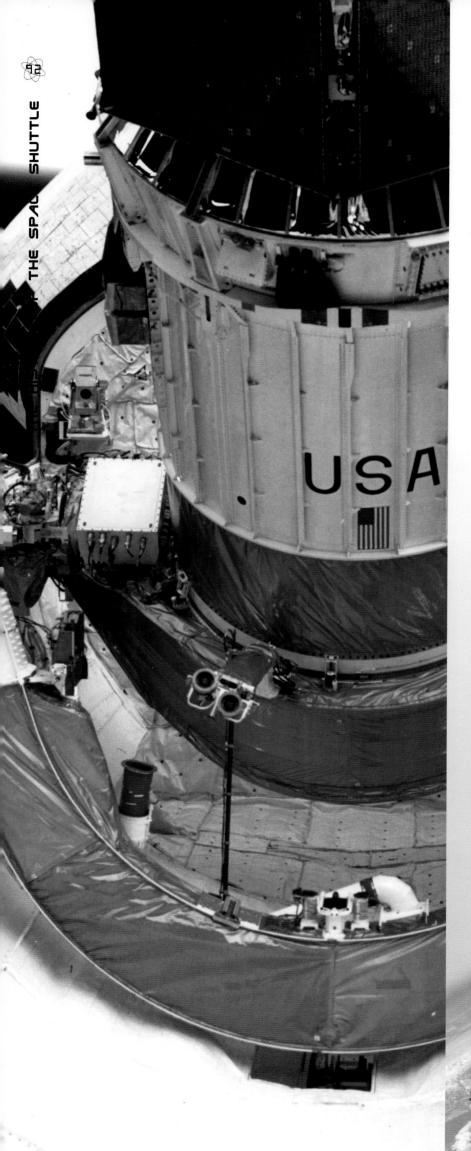

# "We Deliver"

One of the primary uses for the space shuttle was the delivery of a wide range of payloads into orbit, so much so that early shuttle crews took up the mantra of "We Deliver" as an unofficial mission slogan.

Early shuttle "deliveries" were mostly military and communications satellites. The first two were launched with STS-5 in 1982 — SBS-3, owned by Satellite Business Systems, and Anik C3, owned by Telesat Canada.

"We worked with the satellite developers, we got to know how the satellites were put together, and we understood thoroughly what they needed for successful deployment," said mission specialist Joe Allen. "That was the first priority of the mission, to deploy successfully the first hardware put into orbit by the space shuttle."

Two missions later, Sally Ride — in addition to being the first American woman in space — became the first astronaut to retrieve a free-flying satellite, snagging SPAS-01 with the shuttle's Remote Manipulator Arm and returning it to the payload bay.

"I remember thinking, 'Oh, my gosh. This is real metal that will hit real metal if I miss. What if we don't capture this satellite? It was easy in the simulators, but is it going to be as easy in orbit?'" recalled Ride.

Early shuttle crews had plenty of practice. Between STS-5 and STS-61-C (14 missions), NASA averaged almost three deploys per launch, with some crews doing four apiece (STS-41-B and STS-51-G). Some of those missions were tasked

(Preceding page, far left) During STS-26 in 1988, the Tracking Data and Relay Satellite made the journey into space, as seen from *Discovery*'s aft flight deck windows. (Below) On STS-51D in April 1985, an improvised "fly swatter" is used by the Canadarm to activate the Syncom satellite.

## Hear the STS-51G crew deploy ArabSat

Use your QR Code-Enabled device to see and hear the sights & sounds of space shuttle history!

with recapturing satellites for repair. The STS-41-C crew successfully repaired the malfunctioning Solar Max satellite, and STS-51A brought two satellites back to Earth for further refurbishing.

"Our mission was to deliver two communication satellites to orbit and two errant satellites to recover," said Allen. "I got several somewhat rude notes from my fellow astronauts underscoring the fact that in delivering the two satellites to orbit and picking two up, that neither Dale nor I was to get these satellites confused. In other words, don't bring home satellites that we'd just taken there."

Some of the shuttle's deliveries were destined for places beyond low Earth orbit. The shuttle was the jumping off point for the *Magellan*, *Galileo* and *Ulysses* interplanetary probes, sent to study Venus, Jupiter and the sun, respectively.

"We knew that the *Galileo* mission, if successful, the spacecraft was going to end up in orbit around Jupiter several years later and then there were going to be several years of data and images sent back," said STS-34 Commander Don Williams. "It was going to be a living, ongoing program, and we got to be a part of it. That was a really unique experience."

(Preceding page) The Spartan 201 satellite, held in the grasp of *Columbia*'s Remote Manipulator System (RMS) arm, is backdropped over white clouds and the blue waters of the Pacific Ocean in December 1997. (Right) Three STS-49 astronauts hold onto the 4.5-ton Intelsat VI satellite after a six-handed "capture" was made minutes earlier in May 1992. Pictured are, left to right, Richard Hieb, Thomas Akers and Pierre Thuot. Thuot stands on the end of the Remote Manipulator System arm, from which he had made two earlier unsuccessful grapple attempts on two-person extravehicular activity sessions. Ground controllers and crewmembers agreed that a third attempt, using three mission specialists in the cargo bay of *Endeavour*, was the effort needed to accomplish the capture feat.

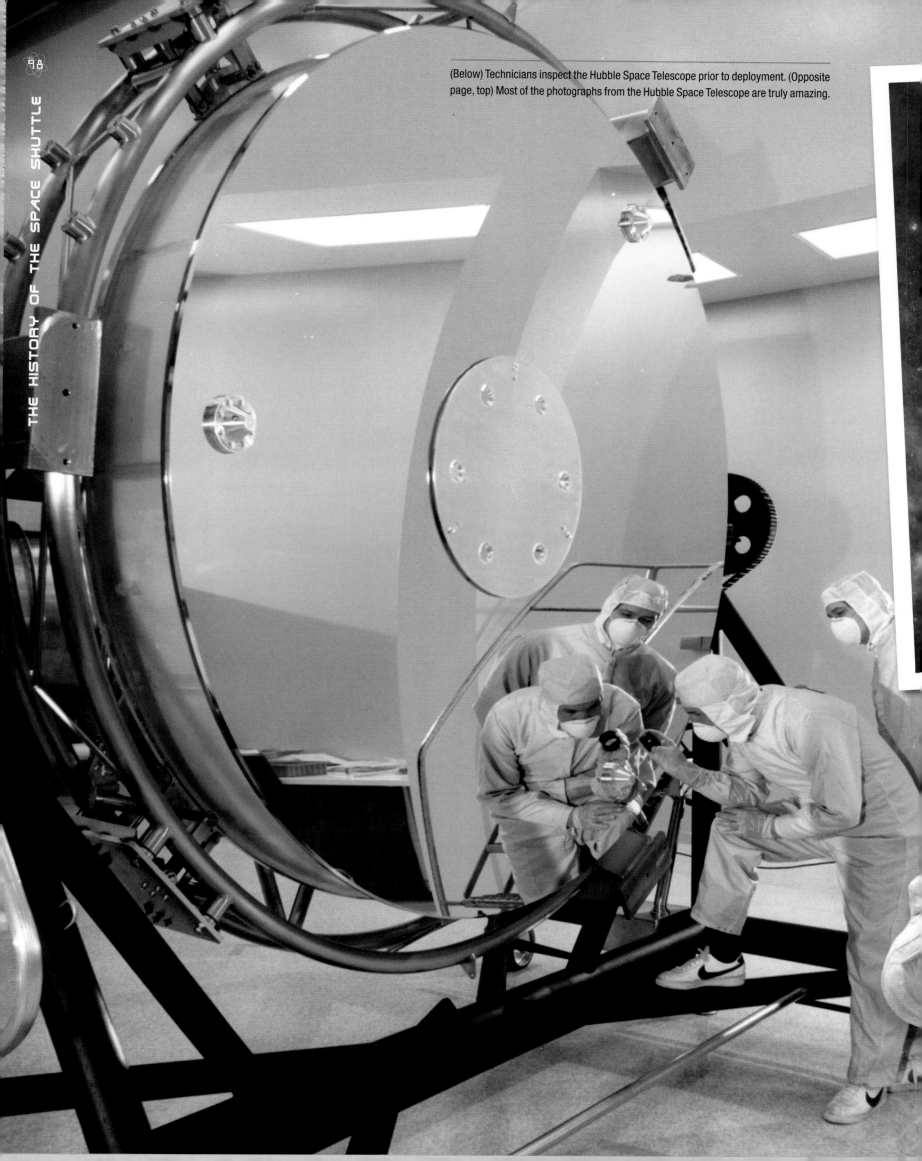

(Below) Technicians inspect the Hubble Space Telescope prior to deployment. (Opposite page, top) Most of the photographs from the Hubble Space Telescope are truly amazing.

NASA flew four more shuttle missions to Hubble to upgrade its capabilities, and to repair and replace worn-out parts, each servicing adding to the functionality and lifespan of this huge astronomical accomplishment.

"I mean, even now, though, in retrospect, I didn't have any idea how significant the discoveries would be and how profoundly revealing the Hubble observations would be," said Hawley. "But I remember thinking, you know, 'This is really going to be special.'"

## Watch STS-125 mission specialist Mike Massimino

interview mission specialist John Grunsfeld about the last Hubble servicing mission

Use your QR Code-Enabled device to see and hear the sights & sounds of space shuttle history!

light, they can be as hot as 250 degrees. A spacesuit protects astronauts from those extreme temperatures.

A spacesuit weighs approximately 280 pounds on the ground — without an astronaut in it. In the microgravity environment of space, a spacesuit weighs nothing.

Putting on a spacesuit takes 45 minutes, including the time it takes to put on the special undergarments that help keep astronauts cool. After putting on the spacesuit, to adapt to the lower pressure maintained in the suit, the astronaut must spend a little more than an hour breathing pure oxygen before going outside the pressurized module.

Shuttle spacesuit materials include ortho-fabric, aluminized mylar, neoprene-coated nylon, dacron, urethane-coated nylon, tricot, nylon/spandex, stainless steel and high-strength composite materials.

Just before a shuttle mission, the suits designated for flight are tested, cleaned and packed at NASA's Johnson Space Center in Houston. Then they are flown to NASA's Kennedy Space Center in Florida and stowed on the shuttle orbiter. After each flight, the suits are returned to Johnson for post-flight processing and reuse.

The longest EVA was 8 hours and 56 minutes, performed by Susan J. Helms and James S. Voss during STS-102 on March 11, 2001.

The first EVA where an astronaut performed an in-flight repair of the space shuttle orbiter was by American astronaut Steve Robinson on Aug. 3, 2005, during STS-114. Robinson removed two protruding gap fillers from space shuttle *Discovery*'s heatshield while the shuttle was docked to the ISS.

Cosmonaut Anatoly Solovyev holds the record for the most spacewalks with 16, with a total duration of 82 hours and 22 minutes, while Michael Lopez-Alegria holds the American record for number of EVAs with 10, with a total duration of 67 hours and 40 minutes.

Of course, you need to spacesuit to walk in space, and the suits are pretty amazing themselves. In Earth orbit, conditions can be as cold as minus-250 degrees Fahrenheit. In the sun-

(Above) STS-82 mission specialist Mark Lee logged 19 hours and 10 minutes on three EVAs during the mission. (Right) STS-61 mission specialist Kathryn Thorton takes part in the first Hubble servicing mission in December 1993. (Opposite page) F. Story Musgrave does some maintenance work on the Hubble during STS-61.

**Hear STS-109's John Grunsfeld discuss** the fit of an EVA suit

Use your QR Code-Enabled device to see and hear the sights & sounds of space shuttle history!

# Laboratory In Orbit

**W**hile many of the space shuttle's flights were tasked with delivering hardware into orbit, almost all of them carried experiments in a variety of scientific fields, such as atmospheric and plasma physics, astronomy, solar physics, material sciences, technology, life sciences and Earth observations.

Some launches, like the STS-9 Spacelab mission, were devoted wholly to scientific endeavors. One of the most interesting experiments on that flight was the growing of protein crystals in zero gravity. Scientists found that crystals grown in space are larger and more neatly ordered — and thus easier to subject to X-ray structural analysis — than those grown here on Earth. Space-grown crystals therefore have great potential to help scientists understand how certain proteins work, perhaps leading to better and more targeted drugs in the future.

(Above) Mission specialist Charles "Sam" Gemar works with the Middeck Zero-Gravity Dynamic Experiment on STS-62. (Below) Eye exams are hardly routine on the space shuttle. (Opposite page) Astronauts Kathryn Sullivan and Charles Bolden work on an ion arc test aboard STS-31 for Utah State student Greg Peterson, who was monitoring the experiment from Mission Control in Houston.

**Hear astronaut Eileen Collins** talk about the experiments aboard STS-93

Use your QR Code-Enabled device to see and hear the sights & sounds of space shuttle history!

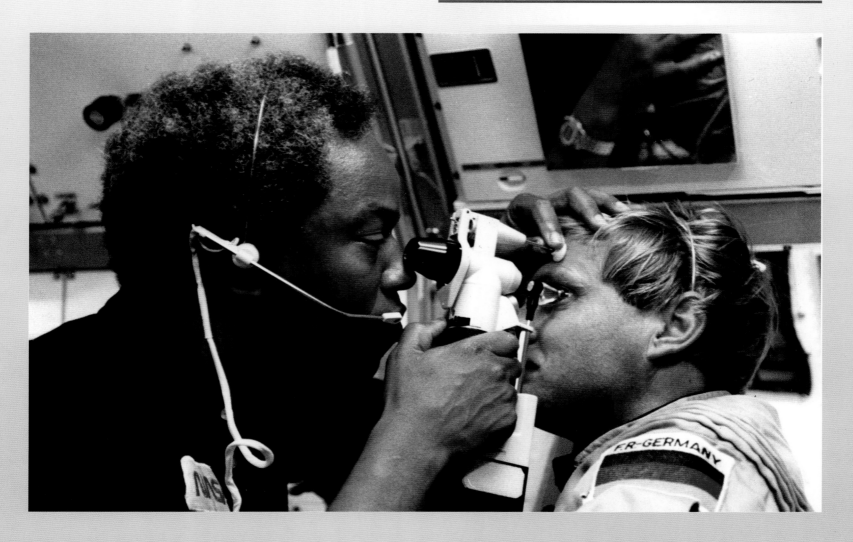

(Above) Michael Good, STS-125 mission specialist, is busy unpacking equipment on *Atlantis'* middeck during his second day in space in preparation for numerous experiments.

Crystals aren't the only thing to react differently in microgravity. Samples of fire moss that traveled onboard the STS-87 space shuttle grew in strange spirals. On Earth, moss spores that take root send out hundreds of tiny filaments known as protonemata. These filaments normally grow in an unruly fashion; they make a messy-looking tangle. But the moss onboard *Columbia* formed a distinctive clockwise spiral.

Even smells can be different in space. On STS-95, the shuttle *Discovery* carried a single miniature rose plant into orbit. Astronauts sampled its volatile oils, which is the substance that carries the essential odors of the flower, and found that the space rose produced fewer volatiles than its coun-

terparts did back on Earth. But, more importantly, its overall fragrance was entirely different. The International Flavors and Fragrances company sampled the space-grown rose and has incorporated the scent into a commercially available perfume.

Being in space affects living organisms as well. On *Atlantis'* STS-115 mission, scientists noticed that samples of Salmonella bacteria got more virulent in space. The findings were confirmed on the STS-123 mission two years later. Salmonella becomes three to seven times more virulent in microgravity conditions because spaceflight tricks them into behaving as if they are inside the human stomach.

Spacelab J, a joint venture between NASA and the National Space Development Agency of Japan that flew aboard STS-47, did a number of life sciences experiments, including several on payload specialist Mamoru Mohri. Another experiment used pairs of small fish to help determine the cause of space motion sickness, which can be debilitating and even dangerous to astronauts in some situations. One set of fish had their otoliths — structures in the inner ear that help animals with balance and orientation — removed, and the other set did not. Since the otoliths are free from gravity in orbit, they do not give the same cues to the brain for certain orientations and motions. It was speculated that the disconnect between the otoliths and what other organs, such as the eyes, were experiencing, could be the cause of space sickness.

During its 30 years in service the space shuttle has carried thousands of experiments into orbit, some for long-duration stays on the International Space Station, and it's certainly safe to say the shuttle has earned a place in the history of scientific discovery.

(Right) STS-73 marked the second flight of U.S. Microgravity Laboratory (USML) and built on the foundation of its predecessor, which flew on *Columbia* during STS-50 in 1992. (Below) U.S. Sen. John Glenn volunteered for numerous experiments on STS-95, including wearing sleep-monitoring devices.

**Hear astronaut Robert Curbeam** talk about a controlled fire experiment on STS-98

Use your QR Code-Enabled device to see and hear the sights & sounds of space shuttle history!

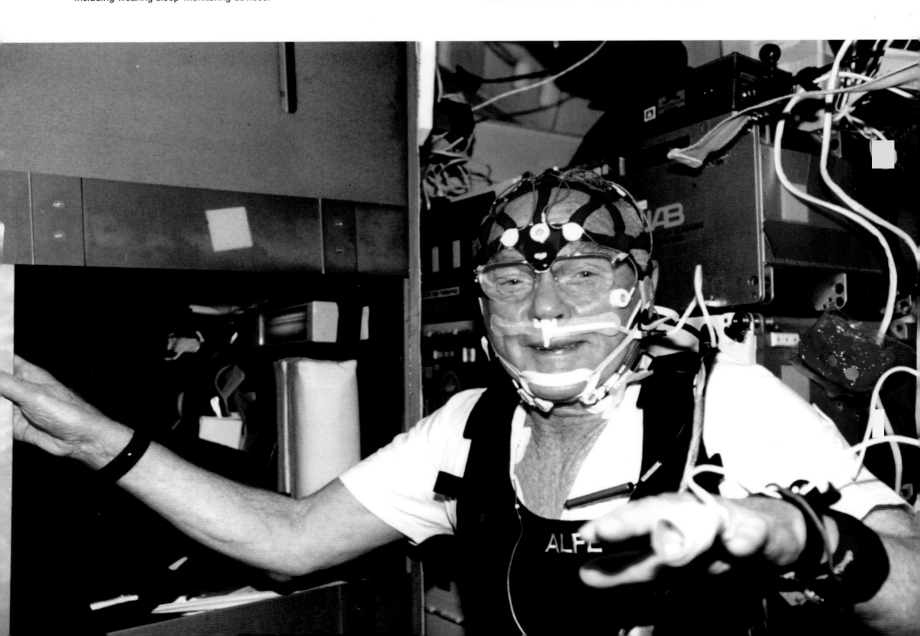

# THE BRICK MOON PROPHECY

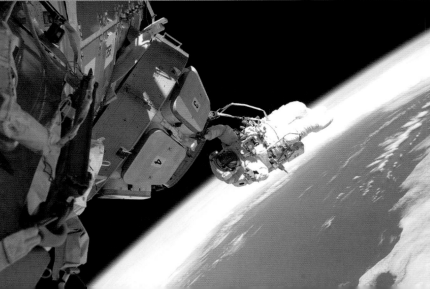

In 1869, clergyman Edward Everett Hale became the first person to imagine a manned station in space in his short novel *Brick Moon*. In 1923, Transylvanian mathematics professor Hermann Oberth published *The Rocket into Interplanetary Space*, and was the first to coin the term "space station" to describe an orbiting facility for staging and refueling voyages to distant planets. Nearly 30 years later, rocket scientist Wernher von Braun published a series of fanciful articles in *Collier's* magazine describing a wheel-like space station that could orbit more than 1,000 miles above the Earth. Some historians believe these articles inspired President John F. Kennedy to announce his intent of putting a man on the moon during his presidency.

In 1971, just 10 years after sending the first human into space, the Soviet Union launched Salyut 1, the world's first space station. Not to be outdone, the United States sent the larger Skylab into orbit just two years later. Before it was abandoned in 1974, Skylab had housed three crews. In 1984, President Ronald Reagan announced plans to construct a permanently manned station within 10 years, underscoring NASA's plans to seek international participation in the project. Four years later, 11 nations signed an agreement to participate in "Space Station Freedom."

After 40 launches of U.S. space shuttles and Russian rockets, over 100 International Space Station components were sent into space so astronauts could construct the new station with the help of ground-breaking robotic technology. Since Nov. 2, 2000, the ISS has been in continuous service, staffed with crewmembers from across the globe.

(Preceding page, top) STS-116 mission specialist Robert Curbeam Jr. prepares to replace a faulty TV camera on the exterior of the International Space Station in 2006. The ISS has been in continuous service since Nov. 2, 2000, thanks to over 40 launches of U.S. space shuttles and Russian rockets to haul components to space to build the station. (Preceding page, bottom left) STS-115 Commander Brent Jett Jr. helps Joseph Tanner with the helmet for his EMU spacesuit in the Quest Airlock of the ISS while *Atlantis* was docked with the station. (Preceding page, bottom right) Nicholas Patrick performs a spacewalk during STS-130 as *Endeavour* delivers a third connecting module — the Tranquility node — and a seven-windowed cupola to be used as a control room for robotics on the ISS. (Below left) When the windows on the ISS' cupola were opened, astronauts had a bird's-eye view of the Sahara Desert.

## See NASA Associate Administrator

for Space Operations William H. Gerstenmaier talk about the space shuttle's contribution to the International Space Station

Use your QR Code-Enabled device to see and hear the sights & sounds of space shuttle history!

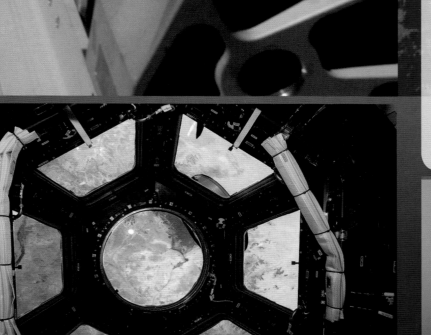

# 15 Years & 1,000 Hours:
# Building the International Space Station

For the past 15 years, the creation of the International Space Station (ISS) has been the primary focus of space programs worldwide.

In 1998, work began in earnest with the robotic launching of Russian live/work module Zarya. Zarya was designed to furnish propulsion and orientation control, electrical power and to additionally serve as a communications hub. Just two weeks later, NASA's space shuttle carried the Unity module to the ISS. Upon arrival, it was attached to Zarya by astronauts using pressurized mating adaptors that allow future space shuttle missions to dock to the station.

(Below) The sign on the fence at Launch Pad 39A announces the mission of STS-88 and *Endeavour* as it sits on the pad. The mission delivered the Unity connecting module to the ISS. (Right) The mated Russian-built Zarya, on left, and U.S.-built Unity modules are backdropped against the blackness of space and Earth's horizon shortly after leaving *Endeavour*'s cargo bay during STS-88.

**Unity – *International Space Station Begins***

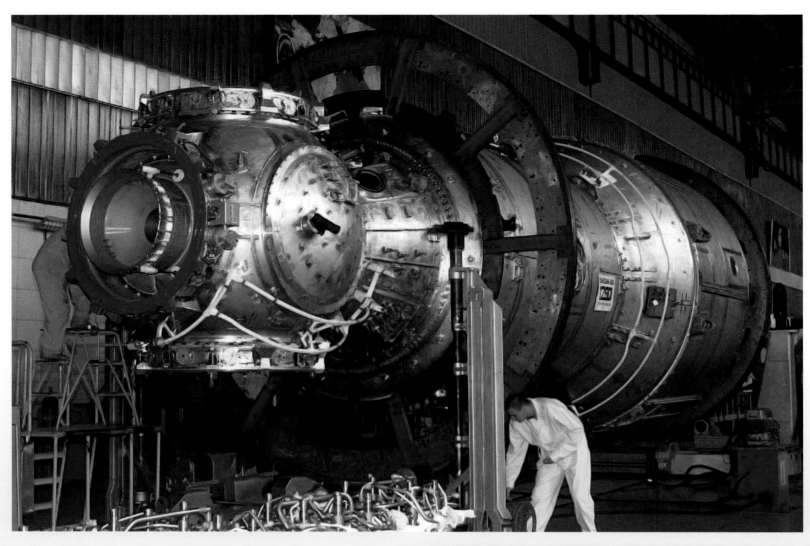

(Above) The Zvezda Service Module, the first Russian contribution and third element to the ISS, is shown under construction in the Krunichev State Research and Production Facility in Moscow. (Right) The Unity connecting module is moved toward the payload bay of *Endeavour* at Launch Pad 39A.

Next in line was the Zvezda module, launched into orbit early in 2000. After Zvezda was connected to the Zarya/Unity modules via ground-control commands, Zarya electronically transferred control of the station to the Zvezda's state-of-the-art computer. The new module added much-needed living quarters, sanitary and kitchen facilities, dehumidifier and oxygen-filtration systems. The additional inclusion of data, voice and television communications with Earth finally made the ISS ready for permanent occupation. In November 2000, the crew of Expedition One became the first to take up residency, adding more advanced television communication systems and supplementary solar panels.

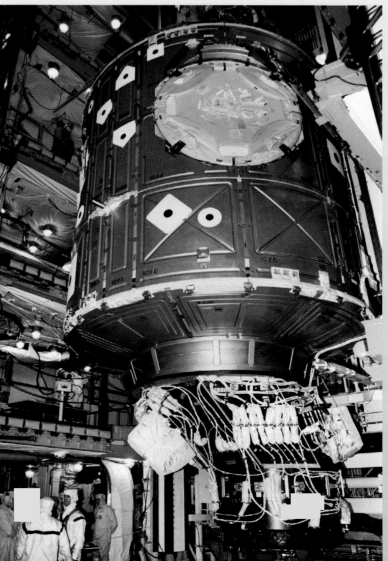

**Hear STS-88 crewmembers talk** about the ISS as a stepping stone

Use your QR Code-Enabled device to see and hear the sights & sounds of space shuttle history!

Expansion continued between 2001 and 2002 when a Russian rocket delivered the Pirs docking compartment. Later, the Destiny laboratory, an airlock system and Canadarm2 (the station's primary robotic arm) were delivered by the shuttles *Discovery*, *Atlantis* and *Endeavour*. The tragic loss of the space shuttle *Columbia* in 2003 halted all station construction for the next two years.

When *Atlantis* arrived at the station in 2006, construction was finally resumed with the delivery of a new set of solar panels. The next three missions delivered more panels, which strengthened ISS power plant to the extent that it could now accommodate additional pressurized modules. Late in 2007, Harmony and Columbus were added, providing much-needed additional laboratory and work space. Next, the first two pieces of the Japanese Experiment Module (or Kibō) were sent into orbit. When the third and last segment was added in 2009, Kibō comprised the largest module of the ISS.

Tranquility was delivered to the ISS via shuttle early in 2010 along with the large-window module cupola. Similar to the Mir Module sent into orbit in 1995, Russia's Mini-Research Module (or Rassvet) is primarily utilized for storage and to provide additional docking ports if needed. The last pressurized module to be sent to the ISS was Leonardo. Sent into orbit aboard the space shuttle *Discovery*, Leonardo was designed to provide crew members with a central place to store supplies and waste products.

Currently, the ISS boasts 15 pressurized modules. Still to be launched, however, is Russia's Multipurpose Laboratory Module. Also called Nauka, the node was originally planned for docking and stowage, but now will be used to replace the aging Pirs. Lastly, the European Robotic Arm will be attached to work on the Russian space station segments currently served by the cranes attached to Pirs. NASA estimates that when construction is completed in 2012, the entire ISS structure will weigh in excess of 400 tons.

---

(Preceding page) The last pressurized module to be sent to the ISS was Leonardo aboard *Discovery*. (Below) With a portion of Unity in the foreground, the Zarya Control Module approaches the U.S.-built connecting module. Using *Endeavour*'s 50-foot-long robotic arm, Nancy Currie plucked Zarya out of orbit.

**Watch STS-88 mission specialist Jerry Ross** talk about the international aspects of the space station:

Use your QR Code-Enabled device to see and hear the sights & sounds of space shuttle history!

# Keeping
# **The ISS**
## in Orbit

From its conception to its present-day status, the International Space Station (ISS) has come to represent one of mankind's crowning technological achievements. Untold man hours have been spent in design, construction and assembly, but an equal amount of time and money have been allocated for repairs and maintenance. During the course of its 20-year history, unforeseen issues and mechanical breakdowns have adversely affected the station's carefully planned construction calendar and work schedules. These malfunctions have sometimes forced crewmembers to operate under less-than-optimal working conditions as they struggled to prevent the ultimate mission failure — the forced abandonment of the station.

A few of the Space Station's more notable repairs included:

**2003 — Accumulation of waste after *Columbia* disaster:** After the space shuttle *Columbia* tragedy on Feb. 1, 2003, NASA suspended the shuttle program for nearly three years. This resulted in a great accumulation of waste that adversely affected the station's standard operations for the next two years. It wasn't until July 2005 that *Discovery* arrived at the ISS and worked to eradicate the problem.

(Top) This 2005 scene in Zarya, the functional cargo block for the ISS, serves witness to the transfer of water and supplies to the space station. Cosmonaut Sergei Krikalev can be seen at the far end of the cluttered hallway. (Above) An airlock serves as a temporary storage area for supplies being transferred to the ISS from *Discovery* during STS-114. (Right) C. Michael Foale, Expedition 8 commander and NASA ISS science officer, performs in-flight maintenance on the nadir window in the Destiny laboratory of the ISS. (Opposite page, top) STS-120 pilot George Zamka holds "cufflink" apparatus that was to be attached to damaged solar arrays and take the structural load off of a broken hinge while *Discovery* was docked with the ISS. (Opposite page, bottom) STS-120 mission specialist Scott Parazynski assesses his repair work as the solar array is fully deployed. (Opposite page, inset) The two-foot tear in solar array material that STS-120 crewmembers had to repair.

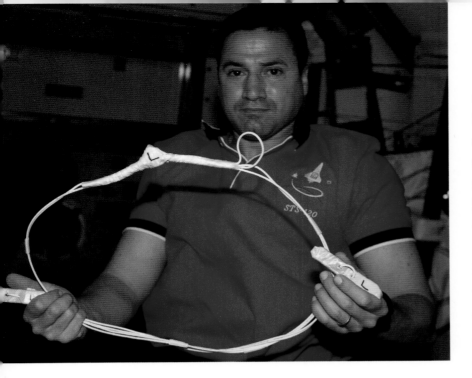

**2004 — Air leak detected:** In January 2004, it was discovered that 5 pounds of air per day were escaping from the ISS. This problem did not cause the crewmembers immediate alarm, but they knew it was something that had to be addressed right away. The leak was difficult to locate due to the day-to-day sounds emitting from the station's equipment, but after a few days, it was discovered on a hose connected to the station's large multi-pane window.

**2007 — Torn solar panel:** During STS-120's visit to ISS, crewmembers repositioned the P6 truss segment, but when they began the deployment of the two solar arrays on the truss, two tears in the photovoltaic blankets were discovered. The mission's spacewalks were replanned in order to devise a repair, and spacewalker Scott Parazynski, assisted by Douglas Wheelock, fixed the torn panels using makeshift cufflinks and riding on the end of the space shuttle's OBSS inspection arm. The spacewalk was significantly more dangerous because of the possibility of shock from the electricity generating solar arrays, the unprecedented usage of the OBSS, and the lack of spacewalk planning and training for the impromptu procedure.

## Watch an animation of all of the upgrades
made to the International Space Station

Use your QR Code-Enabled device to see and hear the sights & sounds of space shuttle history!

# Hooking Up To
# The ISS

Almost all of the space shuttle's last 40 missions involved docking with the International Space Station, and shuttle astronauts train hard for this critical phase of the mission. When guiding a 200,000-pound object moving at more than 17,000 miles per hour toward a seven-story structure orbiting 220 miles above the Earth, any mistake can have disastrous and even deadly consequences.

The system that the shuttle uses to dock with the station has three parts: the external airlock, truss assembly and the androgynous peripheral docking ring. The airlock, located in the payload bay, provides an airtight internal tunnel between the two spacecraft after docking. The truss assembly provides a sound structural base within which the components of the docking system are housed. The truss assembly is physically attached to the payload bay and houses rendezvous and docking aids, such as camera/light assemblies and trajectory control systems. The docking ring consists of a structural base ring housing 12 pairs of hooks, an extendable guide ring with three petals and a motor-driven capture latch within each guide petal. Two control panels in the aft flight deck and nine avionics boxes in the subfloor of the external airlock provide power and logic control of the mechanical components.

**Watch a video of the STS-134 crew** being greeted by ISS astronauts after docking

Use your QR Code-Enabled device to see and hear the sights & sounds of space shuttle history!

(Preceding page top and this page) This image of the International Space Station and the docked *Endeavour*, flying at an altitude of approximately 220 miles, was taken by Expedition 27 crewmember Paolo Nespoli from the Soyuz TMA-20 following its undocking May 23, 2011. The photos taken by Nespoli are the first taken of a shuttle docked to the ISS from the perspective of a Russian Soyuz spacecraft. (Preceding page bottom) This view of *Endeavour*'s forward section — including a partial view of the crew cabin, forward payload bay, docking mechanism and Canadarm — was provided by an Expedition 27 crewmember during a survey of the approaching STS-134 vehicle prior to docking with the ISS. As part of the survey and part of every mission's activities, *Endeavour* performed a backflip for the rendezvous pitch maneuver. (Inset below) As seen through a window on *Endeavour*'s aft flight deck, the ISS affords cosmonaut Vladimir Dezhurov, left, and STS-108 pilot Mark Kelly a farewell look from the shuttle following undocking.

**Watch a video of astronaut Leland Melvin**

talking about having dinner aboard the ISS

Use your QR Code-Enabled device to see and hear the sights & sounds of space shuttle history!

(Above) STS-127 mission specialist Christopher Cassidy uses a range-finding device on *Endeavour*'s flight deck to determine the distance between the shuttle and the ISS during rendezvous and docking activities. (Opposite page, top left) This closeup view of the Pressurized Mating Adapter 2 was taken from *Discovery*'s cabin shortly after docking. (Opposite page, middle right) STS-96 Mission Commander Kent Rominger is about to dock *Discovery* to the ISS. Rominger is at the shuttle's controls on the aft flight deck. The docking mechanism of the approaching station is just a few meters away on the other side of the overhead window. (Opposite page, bottom) Expedition 21 and STS-129 crewmembers gather for a meal at the galley in the Unity node of the ISS while *Atlantis* remains docked with the station. Pictured are Frank DeWinne, Charles Hobaugh, Mike Foreman, Randy Bresnik, Leland Melvin and Robert Satcher Jr.

The docking system is launched with the active docking ring fully retracted and aligned in its final position, the structural hooks open, and the capture latches closed. In preparation for docking, the ring must be extended to its ready-to-dock or ring initial position. The ring is usually moved into the initial position two or three days before the actual docking.

During the rendezvous maneuvers the top of the orbiter is pointed at the other spacecraft. This allows the orbiter to use radar or laser to determine the distance to the other spacecraft. There are a couple of stations keeping positions before the actual docking. The last position is only 30 feet away. The final rate of closure rate is about 0.1 foot/second.

Once capture is achieved, the automatic docking sequence is initiated. After a five-second delay, three electromagnetic brakes are energized for 30 seconds to damp relative motion. Sixty seconds after capture, the ring will start to drive out. The crew will then stop the ring drive. The crew will wait up to eight minutes to allow relative motion to damp.

After approximately three minutes of ring retraction, the ring will activate the ready-to-hook sensors. The ready signal activates close commands and the hooks begin driving closed. As the hooks drive closed, the mating surfaces will compress the pressure seals. Once either set of hooks is closed, the ring is extended slightly to relieve loads on the capture latches. The capture latches are opened and the ring is retracted to its final position. This completes the docking.

The vestibule, the passage between the orbiter and International Space Station, is pressurized and leak checked before the hatches are opened to allow crew and payload transfer.

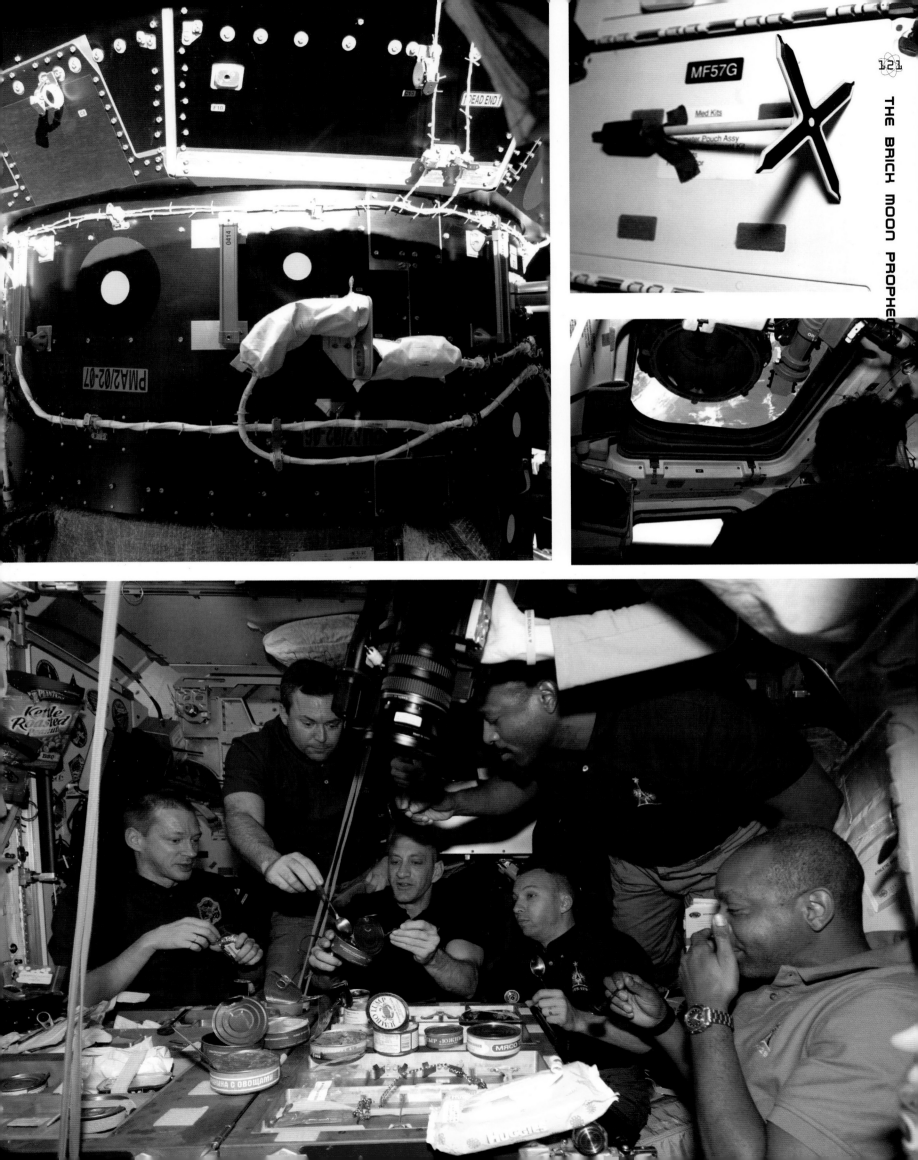

# Getting Supplies & Equipment to the ISS

**Hear STS-108 Commander Dominic Gorie** talk about maneuvering the shuttle and ISS together

Use your QR Code-Enabled device to see and hear the sights & sounds of space shuttle history

Atlantis

For decades, NASA's space shuttle has provided tons of supplies and materials to crewmembers serving on the International Space Station (ISS). However, for over 30 years, unmanned supply craft have also provisioned crews in orbit, supplying Salyut 6 and Salyut 7, Mir and most recently the ISS. Not only do these robotic spacecraft keep operational costs to a minimum, they don't require the life-support systems necessary on the U.S. space shuttles. This important space-saving feature allows for the addition of extra cargo and supplies.

The U.S. was the only country sending manned shuttles into orbit to visit the ISS. But without the aid of Russian, European and Japanese supply spacecraft, providing the necessary items for survival to space station crews would not have been possible.

From the late 1970s, the Russians have been utilizing *Progress*, an expendable freighter spacecraft that is launched unmanned via a Soyuz rocket. After it docks with the space station, astronauts can enter *Progress* for easier removal of cargo. Originally developed to supply the space stations of

(Preceding page) A nadir view of *Atlantis* and its payload. Seen at the rear of the cargo bay is the Raffaello Multi-Purpose Logistics Module, packed with supplies and spare parts for the ISS. (Below) STS-91 mission specialist Wendy Lawrence is almost lost amidst supplies and equipment to be transferred from *Discovery* to the Mir space station. (Left) Philippe Perrin, Yury Onufrienko, Kenneth Cockrell and Daniel Bursch work in close quarters on the middeck of *Endeavour* during STS-111. The limited space was a temporary issue after supplies were moved onto the ISS. (Below left) STS-102's James Kelly, left, and James Wetherbee participate in the movement of supplies inside Leonardo, the Italian Space Agency-built Multi-Purpose Logistics Module.

the Soviet Union, it is currently used to serve the ISS. *Progress* visits the space station three to four times a year and remains docked to the ISS until the next craft arrives with new supplies. Before it is disconnected from the station, *Progress* is then filled with various waste products loaded onboard by the crew and then cast adrift. Soon after, it is completely destroyed upon re-entry into the Earth's atmosphere.

In 2006, the European Space Agency (ESA) began developing plans for an Automated Transfer Vehicle (ATV). Similar to the Russian *Progress* freighters, the ATVs (named *Jules Verne* and *Johannes Kepler*) were created to provision station crewmembers. They also serve as a holding tank for the station's waste products. But unlike *Progress*, the 20-ton ATVs have over three times the capacity and can transport up to eight tons of equipment and cargo. Once docked to the ISS, astronauts can enter the freighter's "shirt-sleeve" environment without the necessity of having to don a space suit. At 10- to 45-day intervals, the ATV's thrusters are utilized to boost the station's altitude. At the end of its mission, the ATV separates from the ISS and is sent into a controlled destructive re-entry into the Earth's atmosphere many miles over the Pacific Ocean.

(Top left) Michael "Rich" Clifford checks over stowed bags filled with extravehicular activity supplies in the STS-76 tunnel. (Middle left) STS-110 Mission Commander Michael Bloomfield moves a water container in the Destiny laboratory on the ISS. (Bottom left) John L. Phillips, Expedition 11 NASA Space Station science officer and flight engineer, participates in the movement of supplies and equipment inside Raffaello. (Below) STS-126 mission specialist Heidemarie Stefanyshyn-Piper works with the transfer of supplies in the Columbus lab of the ISS.

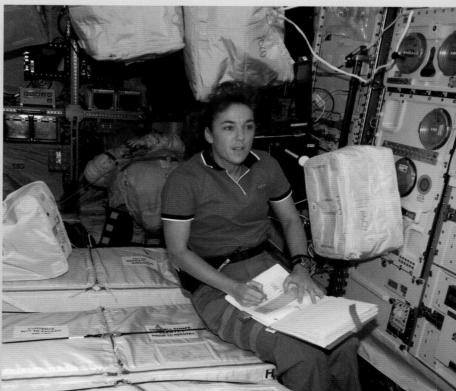

Since the early 1990s, the Japan Aerospace Exploration Agency (JAXA) was steadily working on the design and construction of their H-1 and H-II Transfer Vehicles (HTV). Called *Kibo* and *Kounotori* respectively, *Kibo* launched Sept. 10, 2009. Like the ATV, the HTV carries more than twice the freight payload of the *Progress* but it is only launched approximately once a year. HTVs do not dock automatically like their *Progress* and ATV counterparts, but approach the ISS based upon crew or ground signals. Once they are near the assigned berthing port, robotic arms reach out and pull the HTV into place. Next, International Standard Payload Racks (ISPRs) are moved via computer command into the HTV's main compartment. Like the ATV, HTVs are designed with an easy-access, shirt-sleeve environment in mind.

(Right) Sergei Krikalev of Russia's Federal Space Agency and commander for Expedition 11 retrieves supplies from Raffaello, which was delivered to Earth's orbit by the seven-member crew of *Discovery*. (Below) Scott Altman moves through the tunnel to the ISS with a new battery in hand. The seven-man STS-106 crew experienced a major moving of supplies and hardware from *Atlantis* to the station. (Bottom right) STS-106 mission specialist Edward T. Lu handles part of a treadmill device on the Service Module headed for the ISS.

**Hear STS-112 mission specialist Sandy Magnus** talk about transferring cargo from the shuttle to the space station

Use your QR Code-Enabled device to see and hear the sights & sounds of space shuttle history!

FRANCIS "DICK" SCOBEE
MICHAEL J SMITH
ELLISON S ONIZUKA
JUDITH A RESNIK

RONALD E MCNAIR
S CHRISTA MCAULIFFE
GREGORY B JARVIS

# SACRIFICE

Prior to the start of the space shuttle program, the United States had very few fatal accidents involving astronauts in training for missions, and most of them involved the T-38 supersonic jet aircraft during training flights, but never during an actual spaceflight.

Three astronauts were killed in *Apollo 1* during a command module test on the launch pad at Cape Canaveral on Jan. 27, 1967, due to a cabin fire: command pilot Virgil "Gus" Grissom, senior pilot Edward White and pilot Roger Chaffee. Safety and accountability under the terms "Tough and Competent" was re-emphasized throughout NASA, and modifications were made to Apollo vehicles.

Outside of the harrowing safe flight home of *Apollo 13* from the moon's orbit after an explosion on the outside of the service module that powered the craft, subsequent missions were free of catastrophic outcomes.

That changed Jan. 28, 1986, when the shuttle orbiter *Challenger* broke apart over the Atlantic Ocean after an explosion 73 seconds after launch from Cape Canaveral. The tragedy that took the lives of seven crewmembers on STS-51L highlighted the dangerous nature of spaceflight, and the shuttle fleet was grounded for almost three years as changes were made throughout NASA and the shuttle program.

The shuttles had mostly uneventful flights until Feb. 1, 2003, when the orbiter *Columbia* disintegrated during re-entry over Texas and Louisiana, killing all seven crewmembers on STS-107, and once again the nation mourned the ultimate sacrifice of its astronauts.

The shuttles returned to service in July 2005, and the fleet's last 22 flights were completed safely.

(Preceding page, top left) Crewmembers of STS-51L stand in the White Room at Pad 39B. From left are Sharon "Christa" McAuliffe, Gregory Jarvis, Judy Resnik, Dick Scobee, Ronald McNair, Michael Smith and Ellison Onizuka. The crew perished aboard *Challenger* on Jan. 28, 1986. (Preceding page, middle) On Feb. 1, 2003, seven STS-107 crewmembers died aboard *Columbia*. Seated in front, from left, are Rick Husband, Kalpana Chawla and William McCool. Standing are David Brown, Laurel Clark, Michael Anderson and Ilan Ramon. (Preceding page, bottom and right) Memorials were established to honor the fallen astronauts. (Above) A wing from *Challenger* was among the recovered debris.

**Hear STS-88 Commander Robert Cabana** talk about the risks of exploring space

Use your QR Code-Enabled device to see and hear the sights & sounds of space shuttle history!

# "Slipped The Surly Bonds Of Earth"

**T**he liftoff of space shuttle mission STS-51L on Jan. 28, 1986, was intended to be like the 24 others taken since the first in April 1981 from Cape Canaveral. Crew sizes had increased over the years, and this particular mission had the first civilian, schoolteacher Christa McAuliffe, on board the orbiter *Challenger* as part of the "Teacher in Space Project" started by NASA.

As schoolchildren across the nation watched on televisions in their classrooms with their own teachers, something went catastrophically wrong with the fuel tanks that surrounded *Challenger* during the craft's race for Earth orbit, and the shuttle broke apart in an explosive plume 48,000 feet above the Earth's surface.

Seven crewmembers were killed in the accident: Commander Francis Scobee, pilot Michael Smith, mission specialists Ellison Onizuka, Ronald McNair and Judith Resnik, and payload specialists Gregory Jarvis and McAuliffe.

The mission's launch, originally scheduled for Jan. 22, had been rescheduled twice and scrubbed four other times due to various issues in the days leading up to Jan. 28.

Jan. 28 had dawned very cold with temperatures below freezing. Ice had formed on the launch tower, and there was concern that the rubber O-rings that helped hold together the solid rocket boosters attached to the shuttle would be too brittle for the forces of the launch.

As the morning progressed, work on the ice commenced by launchpad crews, and enough melting was deemed to have had occurred to allow the shuttle to launch at 11:38 a.m.

(Top left) Schoolteacher Christa McAuliffe, second from left in back, was the first civilian on a shuttle mission. (Middle left) Christa McAuliffe gets briefed about her launch helmet. (Bottom left) On the day of *Challenger*'s launch on Jan. 28, 1986, icicles draped the Kennedy Space Center in Florida. The unusually cold weather, beyond the tolerances for which the rubber seals were approved, most likely caused the ruptured O-ring in the right solid rocket booster that triggered an explosion soon after launch. (Opposite page, top) Mission Control spotted a plume coming from *Challenger* just 58 seconds after liftoff, and within 15 seconds, the vehicle and its rockets were in full failure. (Opposite page, bottom left) At about 76 seconds, fragments of the orbiter can be seen tumbling against a background of fire, smoke and vaporized propellants from the external tank. The left solid rocket booster flies rampant while still thrusting. (Opposite page, bottom right) This photograph shows the main engines and solid rocket booster exhaust plumes entwined around a ball of gas from the external tank.

## Hear the closing words of President Reagan's address after the *Challenger* accident

Use your QR Code-Enabled device to see and hear the sights & sounds of space shuttle history!

Just over 58 seconds after liftoff, a plume could be seen on the right of the solid rocket booster. In a matter of 15 seconds, the vehicle and its rockets were in full failure, and pieces of the orbiter streamed to the ocean below.

NASA public affairs officer Steve Nesbitt said, "Flight controllers here looking very carefully at the situation. Obviously a major malfunction. We have no downlink." Moments later came the confirmation of what everyone had seen. "We have a report from the Flight Dynamics Officer that the vehicle has exploded."

Mission Control in Houston was sealed off from the outside world, and those inside worked on capturing and preserving the data surrounding the accident, and NASA began the grim task of recovery over the next several weeks.

The night of the tragedy, President Ronald Reagan was supposed to deliver his State of the Union speech. Instead, he addressed the nation from the Oval Office about the *Challenger* incident.

In his address, Reagan closed with these words, "We will never forget them, nor the last time we saw them this morning, as they prepared for their journey and waved good-bye, and slipped the surly bonds of Earth to touch the face of God."

Investigations into the cause of the *Challenger* accident were opened soon after, including the Rogers Commission, which included former astronauts Sally Ride and Neil Armstrong. The shuttle fleet was grounded until *Discovery* made its "Return to Flight" mission in September 1988.

**Hear about a tradition** begun during STS-26, the "Return to Flight" after *Challenger*

Use your QR Code-Enabled device to see and hear the sights & sounds of space shuttle history!

(Preceding page, top left) Wreckage from *Challenger* was retrieved from the Atlantic Ocean by a flotilla of U.S. Coast Guard and U.S. Navy vessels. (Preceding page, top right) Search-and-recovery teams located pieces of both the left and right sidewall of *Challenger* during the months-long retrieval effort that followed. (Preceding page, bottom) The debris was impounded at the Kennedy Space Center and the Cape Canaveral Air Force Station. (Above) Entombment of the *Challenger* wreckage took place at abandoned Minuteman missile silos at Complex 31 on Cape Canaveral Air Force Station. (Below) The *Challenger* crewmember remains were transferred from seven hearse vehicles to a MAC C-141 transport plane at the Kennedy Space Center's Shuttle Landing Facility for transport to Dover Air Force Base, Delaware.

# "A High And Noble Purpose"

**N**ASA's space shuttle program, rocked by the *Challenger* tragedy in January 1986, had resumed missions by September 1988, and operations became routine once again as the orbiter was making flights to the Russian Space Station Mir, helping with the construction of the International Space Station and conducting repairs on the Hubble Space Telescope.

STS-107, launched Jan. 16, 2003, had seven astronauts on board the shuttle *Columbia*: Commander Rick Husband, pilot Willie McCool, payload commander Michael Anderson, mission specialists David Brown, Kalpana Chawla and Laurel Clark, and payload specialist Ilan Ramon, a colonel in the Israel Air Force.

The crew was on a 16-day research mission, but all were killed 16 minutes before their scheduled landing Feb. 1 at the Kennedy Space Center in Florida when the orbiter broke up over Texas during re-entry.

The cause of the accident was traced back to the launch at the Kennedy Space Center, when a piece of foam insulation broke off of the shuttle's external tank as the orbiter left the launch pad and struck *Columbia*'s reinforced carbon-carbon heat shield on the left wing's leading edge. The strike damaged the shuttle's thermal protection system, which protects the orbiter from the heat of re-entry.

As a result, hot gases from re-entry penetrated the wing, which caused the orbiter to disintegrate in a catastrophic failure at 9 a.m. Eastern time. Mission Control alerted search-and-rescue teams 12 minutes later after reports of disintegration of *Columbia* by eyewitnesses reached NASA officials.

(Preceding page) After the *Challenger* tragedy in 1986, NASA was ready to return to space by 1988. Missions again became routine until Feb. 1, 2003, when seven astronauts perished on *Columbia* during re-entry (preceding page, bottom right) over Texas. (Top right) After the tragedy, people placed signs, U.S. flags and flowers on the fences near the main entrance at the Johnson Space Center. (Right) President George W. Bush speaks at a memorial for STS-107 crewmembers.

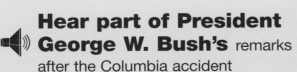 **Hear part of President George W. Bush's** remarks after the Columbia accident

Use your QR Code-Enabled device to see and hear the sights & sounds of space shuttle history!

Later that afternoon, President George W. Bush addressed the nation from the White House Cabinet Room. He said, in part:

"In an age when space flight has come to seem almost routine, it is easy to overlook the dangers of travel by rocket, and the difficulties of navigating the fierce outer atmosphere of the Earth. These astronauts knew the dangers, and they faced them willingly, knowing they had a high and noble purpose in life. Because of their courage and daring and idealism, we will miss them all the more."

A large recovery operation began almost immediately. Over 2,000 debris fields from southeast of Dallas into Louisiana and Arkansas were found, and pieces of the orbiter were sent to NASA facilities for cataloging and examination.

The Columbia Accident Investigation Board was convened, and as it was after the *Challenger* accident, NASA was criticized for its decision-making and risk assessments. A foam-impact test showed a loose piece of insulation could

(Above) Rev. Robert L. Bush, pastor at First Church of the Nazarene in Lufkin, Texas, delivers an invocation at a memorial service honoring the *Columbia* crewmembers. (Right) Workers bring in pieces of *Columbia*'s debris and place them in the grid on the floor of the RLV Hangar.

cause a significant breach of the leading edge of the wings' thermal protection system during launch.

The shuttle program was suspended while the causes of the *Columbia* accident were investigated. The suspension also delayed the construction of the ISS.

The fleet went back into service with the "Return to Flight" launch of the orbiter *Discovery* for STS-114 on July 26, 2005. Foam insulation was shed from the external tank during that launch, but did not strike the orbiter. As a result, the shuttle program was grounded again until July 2006 when missions resumed.

(Above) In the Columbia Debris Hangar, members of the Stafford-Covey Return to Flight Task Group look at recovered tiles. Chairing the task group were former shuttle Commander Richard Covey and former Apollo Commander Thomas Stafford.

**See part of NASA Administrator Sean O'Keefe's address** during the *Columbia* memorial service in Lufkin, Texas

Use your QR Code-Enabled device to see and hear the sights & sounds of space shuttle history!

# FINAL MISSION

**Hear the crew of STS-135,** the final space shuttle mission, talk about the legacy of the shuttle program

Use your QR Code-Enabled device to see and hear the sights & sounds of space shuttle history!

For 30 years, the space shuttle program was the United States' access to space, starting with *Columbia*'s maiden voyage April 12, 1981, with John Young and Robert Crippen aboard. A total of 134 missions later, the shuttle fleet made its final flight as *Atlantis* took to Earth orbit July 21, 2011, on STS-135 from the Kennedy Space Center in Florida.

During the 30 years of the program, a total of 355 individuals — 306 men and 49 women — representing 16 different countries were part of 852 crew slots on shuttle missions flown by five orbiters: *Columbia*, *Challenger*, *Discovery*, *Endeavour* and *Atlantis*. Those astronauts flew a total of over 542 million miles around the Earth.

*Atlantis'* flight marked the emotional end to a program that built a reusable orbiter that could be landed like a commercial jet. Tragedies involving the *Challenger* in 1986 and *Columbia* in 2003 slowed but did not halt the program and its progress.

The final mission was more than a sentimental journey. Like the previous missions, it was a business trip. The crew of Commander Christopher Ferguson, pilot Douglas Hurley, and mission specialists Sandy Magnus and Rex Walheim docked the orbiter to the International Space Station, and they delivered supplies and parts to the ISS. One more payload was released, the 180th by the shuttle.

Finally, there was a triumphant landing at the Kennedy Space Center after 200 orbits. It was the end of an era for an amazing vehicle and a celebration of its legacy in space exploration.

(Preceding page) *Atlantis* flies over the Bahamas prior to docking with the International Space Station on July 10, 2011, during the final space shuttle mission, STS-135. (Below) Attired in training versions of their shuttle launch and entry suits, the STS-135 crew poses for a portrait. From left to right are Rex Walheim, Doug Hurley, Chris Ferguson and Sandy Magnus.

# Launching Into
# History

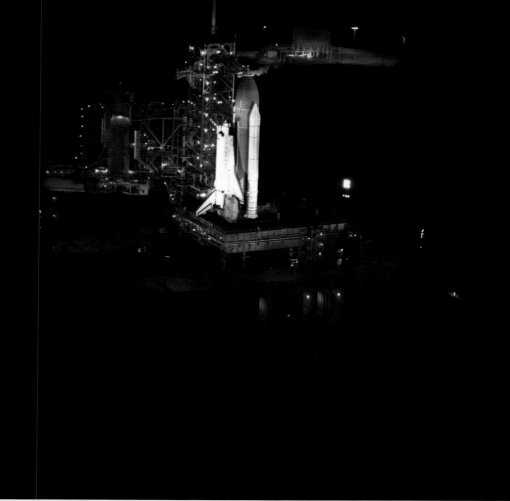

The final launch of the space shuttle fleet July 8, 2011, was a huge event along Florida's east coast. An estimated 1 million spectators lined the area known as Florida's "Space Coast" to watch the liftoff of the orbiter *Atlantis* and its four-person crew of Commander Christopher Ferguson, pilot Douglas Hurley, and mission specialists Sandra Magnus and Rex Walheim.

The crew of four was the smallest to fly on a mission since STS-6 in April 1983. With no shuttles available for a rescue mission if anything went wrong, they would be returned one at a time over a year by way of Russian Soyuz capsules from the International Space Station.

A hot day, in which the weather around the launch pad threatened the 10-minute launch window, dawned at the Kennedy Space Center. Even though there was a 70 percent chance of not being able to launch, the countdown continued as scheduled.

The external fuel tank loading started just after 2 a.m., and two-and-a-half hours later, the 15-story-tall tank was filled with 535,000 gallons of chilled liquid oxygen and liquid hydrogen. Just as the fueling was completed, the four astronauts were awakened and began their own preparations for liftoff.

In their familiar orange spacesuits, the astronauts entered the Astrovan, the silver vehicle that would drive them to Launch Pad 39A, and they began to board *Atlantis* just after

(Above) *Atlantis*, attached to its bright-orange external fuel tank and twin solid rocket boosters, is bathed in xenon lights and takes center stage on Launch Pad 39A at NASA's Kennedy Space Center prior to its final launch. STS-135 was the 33rd flight of *Atlantis*, the 37th shuttle mission to the space station, and the 135th and final mission of NASA's space shuttle program. (Opposite page) The crew acknowledges a crowd prior to liftoff.

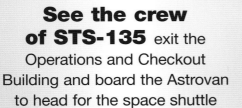 
8 a.m., with Ferguson loading in first, followed by Hurley, then Mangus and finally Walheim. The hatch of the shuttle was shut and sealed by NASA's closeout crew 90 minutes later, and final checks continued.

Two planned holds of the countdown proceeded as scheduled, and everything — including the weather near the Kennedy Space Center — was cooperating as the skies began to clear in the area.

But just 31 seconds to go prior to liftoff, the countdown was held to determine whether the gaseous oxygen vent arm

**See *Atlantis* launch
on the final mission** of
the space shuttle program

Use your QR Code-Enabled device to see and hear the sights & sounds of space shuttle history!

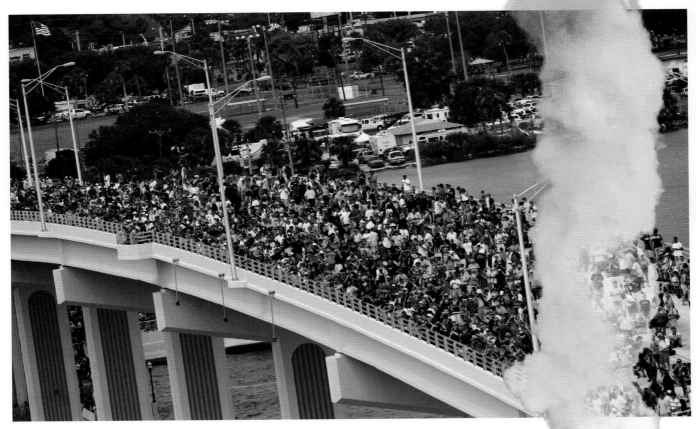

had fully retracted from the shuttle. A video check confirmed the retraction, and the countdown clock was resumed.

With less than a minute left in the launch window, *Atlantis'* main engines and two solid rocket boosters came to life as the countdown headed toward liftoff. The orbiter broke free of Earth as the engines sent the familiar plumes of fire and smoke billowing away from the launch pad at 11:29 a.m. EDT to the cheers of the assembled crowds throughout Florida.

"The space shuttle spreads its wings one final time for the start of a sentimental journey into history," said NASA commentator Rob Navias moments after liftoff as spectators near the launch site lifted their eyes to the heavens as *Atlantis* headed for Earth orbit.

One minute later, *Atlantis* and its crew reached "max-Q," the point of the highest aerodynamic pressure during launch, and a few minutes later, the spent solid rocket boosters fell away from the shuttle toward the Atlantic Ocean and recovery by NASA.

Finally, the shuttle reached Earth orbit to begin its final mission in space.

(Preceding page) Smoke and steam billow outward as *Atlantis* lifts off on twin columns of flame as it begins its STS-135 mission to the ISS. (Above) Residents and visitors to Florida's Space Coast gather on the A. Max Brewer Bridge in Titusville to see the shuttle rockets' red glare for the last time. (Right) The exhaust plume from *Atlantis* as seen through the window of a Shuttle Training Aircraft as STS-135 launches July 8, 2011, in Cape Canaveral, Florida.

# A Supply Ship
# To Last

**E**ven with the hoopla surrounding the final liftoff of the space shuttle fleet, there was a mission ahead for the crew of *Atlantis* as it settled into orbit and begun preparations for docking and work with the onboard crew of the International Space Station.

The Raffaello Multi-Purpose Logistics Module, the robotic refueling mission experiment and the PicoSat satellite were aboard the shuttle for the ride to space. Raffaello was making its fourth trip to the ISS, bringing supplies and returning discarded material back to earth on *Atlantis*.

The crew used the shuttle's robotic arm to check the leading edges of the shuttle and the nose cap over a six-hour period to make sure no damage was done during the launch from Kennedy Space Center, and the crew continued preparations for docking with the ISS.

The following day, *Atlantis* did a slow backflip near the space station to allow ISS crewmembers to take pictures of the shuttle's heat shield for inspection by Mission Control. After that was complete, Commander Christopher Ferguson guided the craft to the ISS, and docked it to the space station. After safety checks, a ceremonial bell was rung in the space station commemorating the arrival of the shuttle crew.

Astronauts moved the Raffaello cargo carrier from the shuttle's cargo bay to a node on the ISS by way of *Atlantis'* robotic arm, the Canadarm2. The hatches were opened between the carrier and the station after the OK was received. The crew transferred 2,300 pounds of experiments, equipment and supplies from *Atlantis'* middeck lockers to the ISS.

(Preceding page, top) Expedition 28 flight engineer Ron Garan used a window on the Zvezda Service Module to photograph *Atlantis*, which can be seen through the window during rendezvous operations on July 10, 2011. (Preceding page, bottom) A reunion of sorts took place among six NASA astronauts, three Russian cosmonauts and a Japanese astronaut in the ISS' U.S. Harmony Node 2 following a July 10, 2011, docking of *Atlantis* and the station. (Below) With his feet secured on a restraint on the Remote Manipulator System's robotic arm, Mike Fossum holds the Robotics Refueling Mission payload, which was the focus of one of the primary chores accomplished on a 6 1/2-hour spacewalk on July 12, 2011.

**See the STS-135 crew** remove the Raffaello Multi-Purpose Logistics Module from *Atlantis'* cargo bay

Use your QR Code-Enabled device to see and hear the sights & sounds of space shuttle history!

(Top left and above) Surrounded by supplies and spare parts in the Raffaello Multi-Purpose Logistics Module, Sandy Magnus works as "load master" for the joint activities of the *Atlantis* and ISS crews. (Top right) STS-135 pilot Doug Hurley tapes up a supply bag on the middeck of *Atlantis* while it was docked with the ISS. (Right) STS-135 Commander Chris Ferguson looks over Raffaello after a great amount of work by joint crews from *Atlantis* and the ISS to transfer supplies to and from the two spacecraft. (Opposite page, top) At a farewell ceremony before the shuttle crew returned to *Atlantis*, STS-135 Commander Chris Ferguson, center with microphone, made special presentations of a U.S. flag and a space shuttle model to the ISS and its current crew. The flag was originally flown into space on the first shuttle mission in 1981. (Opposite page, bottom) Expedition 28 crewmembers and the STS-135 *Atlantis* astronauts form a microgravity circle for a portrait aboard the ISS' Kibo laboratory.

A six-hour-plus spacewalk was conducted the next day to retrieve a failed pump module for return to Earth and to repair a new base for the ISS' robotic arm. After that procedure, work to move the nearly 9,400 pounds of supplies and equipment from Raffaello to the station began with help from ISS crewmembers.

The transfer took a couple of days, and the crew took a call from President Barack Obama, who thanked them and the men and women of NASA for their work, saying, "The space program has always embodied our sense of adventure and explorations and courage."

Once the transfer of supplies from Raffaello was complete, it was time to load 5,700 pounds of discarded materials from the ISS to the cargo carrier, which was returned to the shuttle.

The astronauts provided a recorded message as a tribute to the entire shuttle program. Ferguson talked about the U.S. flag carried with them, the same one flown on STS-1 30 years earlier. The flag will be left on the ISS until the next crew launched from the U.S. retrieves it for return to Earth.

The ISS crew of Commander Andrey Borisenko and flight engineers Mike Fossum, Ron Garan, Satoshi Furukawa, Alexander Samokutyaev and Sergei Volkov went to work stowing the supplies, as *Atlantis* undocked from the station.

*Atlantis'* final payload, the PicoSat, was launched from the craft the next day, and the crewmembers prepared the shuttle for its return to Earth.

**See the crew transferring supplies from *Atlantis*** to the International Space Station

Use your QR Code-Enabled device to see and hear the sights & sounds of space shuttle history!

# Back On **The Ground**

With the supply transfer complete to the International Space Station, and the PicoSat satellite deployed, the crew began their final preparations for landing back in Florida.

Kate Smith's traditional rendition of "God Bless America" woke the crew of *Atlantis* on the final day in space. It was dedicated not to a specific crewmember, as was done on most days of missions, but to the entire crew and "all the men and women who put their heart and soul into the shuttle program for all of these years," said Shannon Lucid, the capsule communicator, said to the shuttle crew.

(Top) Ribbons of steam and smoke trail *Atlantis* as it nears touchdown on the Shuttle Landing Facility's Runway 15 at NASA's Kennedy Space Center for the final time. Securing the space shuttle fleet's place in history, *Atlantis* marked the 26th nighttime landing of a shuttle and the 78th landing at Kennedy. (Above) The drag chute trailing *Atlantis* is illuminated by the xenon lights on Runway 15. (Opposite page, top right) Workers marked in bright red the letters "MLG" at the spot where *Atlantis*' main landing gear came to a stop after the vehicle's final return from space. (Opposite page, bottom) This is a unique look at *Atlantis* on its way home, as photographed by the Expedition 28 crew of the ISS.

Once awake, the crew went to work on getting ready for re-entry of the shuttle, as crews on the ground studied weather reports to see if *Atlantis* could land in Florida as scheduled at 5:56 a.m. EDT. There was a second opportunity, if needed, to touch down about 90 minutes later, or the following day if conditions warranted.

First, the crew stowed the Ku-Band antenna, used for data and television communications from space, and mission managers cleared *Atlantis'* heat shield for use after a final inspection.

Then Mission Control gave *Atlantis* permission to start the de-orbit burn to slow down the orbiter to 223 miles per hour, which allows it to begin its decent into the atmosphere from 240 miles up. The craft is oriented backward, with the three engines facing the direction in which it needs to travel. The burn takes over three minutes, and afterward the shuttle is flipped over to orient it for re-entry. On the ground, lights are lit on the runway to allow the crew to see its target in the darkness.

Commander Christopher Ferguson and pilot Douglas Hurley began to bring *Atlantis* home, passing over the Caribbean Sea and western Cuba before entering Florida airspace. The shuttle crossed over Florida, and it made a sweeping turn over the Atlantic Ocean to align it with the runway at the Kennedy Space Center.

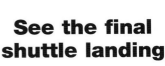 

## See the final shuttle landing

Use your QR Code-Enabled device to see and hear the sights & sounds of space shuttle history!

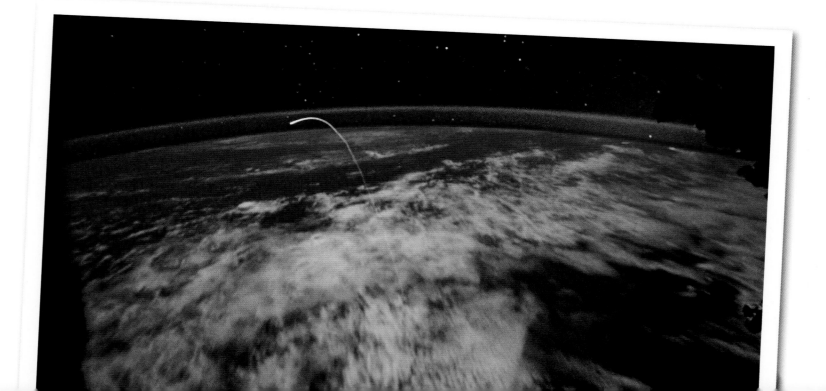

On the ground, twin sonic booms were heard as *Atlantis* approached, and it landed safely at 5:57 a.m., as the drag chute opened behind the spacecraft, slowing it down on the runway. Once on the ground, an arrival checklist was run through onboard the shuttle to shut down the systems.

Ferguson took a moment to say, "The space shuttle changed the way we view the world, the way we view the universe. America's not going to stop exploring. Thanks for protecting us and bringing this program to a fitting end."

Within an hour, support crews assisted the crew from the orbiter, as Ferguson was the last to leave his ship. Later, the crew addressed the media, and Ferguson summed up the mission and the shuttle program with this: "Although we got to take the ride, we hope everyone was able to take a little part of the journey with us."

Ferguson hopes the shuttles on display will inspire a child in the future to say, "Daddy, I want to do something like that when I grow up."

**Hear STS-35 Commander**
Chris Ferguson's remarks after exiting the shuttle for the final time

Use your QR Code-Enabled device to see and hear the sights & sounds of space shuttle history!

(Preceding page, top) The STS-135 crewmembers pause for a final photograph in front of *Atlantis*. (Preceding page bottom, and below) *Atlantis* is rolled over to the Orbiter Processing Facility shortly after completing its 13-day mission to the ISS in 2011. Overall, *Atlantis* spent 307 days in space and traveled nearly 126 million miles during its 33 flights. (Above) STS-135 Commander Chris Ferguson makes a public statement as NASA Administrator Charles Bolden and the three other *Atlantis* astronauts look on after landing at the Kennedy Space Center.

National
# Treasures

HUMAN SPACEFLIGHT

**See NASA Administrator Charles Bolden** announce that the space shuttle *Atlantis* while be on permanent display at Kennedy Space Center

Use your QR Code-Enabled device to see and hear the sights & sounds of space shuttle history

As the wheels of *Atlantis* came to a stop July 21, 2011, at the Kennedy Space Center, the dismantling of the space shuttle program was well under way.

The crew of *Atlantis* was limited to four for the final mission, because any rescue attempt in case of emergency would had have been routed through the International Space Station and Russian Soyuz capsules over a year's time since there was no other active shuttle available for use.

*Discovery* last flew for STS-133, and after the orbiter landed March 9, 2011, back at Kennedy, technicians started deservicing operations on the craft March 29. *Endeavor* was next to finish its service after the STS-134 mission ended with the shuttle's landing at Kennedy on June 1, 2011. The next day, the craft started to be deserviced by NASA technicians. As for *Atlantis*, only four days passed before workers began the long process of stripping the orbiter and preparing it for museum display.

In 2008, NASA had begun the process of deciding where the orbiters would be on display, and that included *Enterprise*, the first-built shuttle that never flew in space. In 2010, the agency requested input from museums and organizations that wished to publically display one of the orbiters at their climate-controlled facility at a cost of $28.8 million.

(Preceding page) The *Enterprise*, the first-built shuttle that never flew in space, was the centerpiece at the Smithsonian's National Air and Space Museum Steven F. Udvar-Hazy Center in Chantilly, Virginia. (Above) It was announced in April 2011 that *Enterprise* would be moved to the Intrepid Sea, Air & Space Museum in New York, while the other orbiters will be on display in Virginia, California and Florida. (Below) NASA Administrator Charlie Bolden, at podium, announces that *Atlantis* will remain on permanent exhibition at the Kennedy Space Center Visitor Complex. To his left is Kennedy Space Center Director Bob Cabana. Also speaking during the ceremony were astronaut Janet Kavandi, USA's Mike Parrish and STS-1 pilot and former KSC Director Bob Crippen.

In April 2011, NASA announced the future homes of the orbiters: *Enterprise* would go from the Smithsonian's National Air and Space Museum Steven F. Udvar-Hazy Center in Virginia to the Intrepid Sea, Air & Space Museum in New York; *Discovery* would be stationed at the Udvar-Hazy Center; *Endeavour* would go to the California Science Center in Los Angeles; and *Atlantis* would have the shortest move, going on display at the Kennedy Space Center Visitor's Complex, not far from where it landed after the end of the final shuttle mission, STS-135.

NASA Administrator Charles Bolden said at the time of the announcement, "In the end, these choices provide the greatest number of people with the best opportunity to share in the history and accomplishments of NASA's remarkable space shuttle program. These facilities we've chosen have a noteworthy legacy of preserving space artifacts and providing outstanding access to U.S. and international visitors."

In addition, NASA sent other artifacts to museums and educational institutions, including: shuttle simulators to Adler Planetarium in Chicago, the Evergreen Aviation & Space Museum in McMinnville, Oregon, and Texas A&M's Aerospace Engineering Department; a full-fuselage trainer for the Museum of Flight in Seattle; and nose cap assembly and crew compartment trainer for the National Museum of the U.S. Air Force at Wright-Patterson Air Force Base in Ohio.

By late September 2011, *Discovery*, *Endeavour* and *Atlantis* were being worked on at the two orbiter processing facilities at Kennedy Space Center. When the work is complete, the crafts will be ferried to the new locations by the same carrier aircraft used during the shuttle program to return the orbiters back to Florida after California mission landings. Once at their new locations, the shuttles will be prepared for permanent display for future generations to marvel at.

## See how NASA plans to display *Atlantis*

Use your QR Code-Enabled device to see and hear the sights & sounds of space shuttle history!

(Preceding page, left) Workers watch as *Atlantis* slowly backs out of Orbiter Processing Facility-1 during its rollover to the Vehicle Assembly Building. (Preceding page, bottom) *Endeavour* and *Discovery* meet nose-to-nose as the vehicles switch locations Aug. 11, 2011, at NASA's Kennedy Space Center. (Top left) In the Orbiter Processing Facility-2, a technician monitors the progress of a special crane as it lifts the forward reaction control system away from *Atlantis*. (Top right) Six space shuttle main engines sit on stands inside the Engine Shop at NASA's Kennedy Space Center where they were being prepared for shipment to the Stennis Space Center in Mississippi for storage following the completion of the shuttle program. The engines will be used on NASA's Space Launch System heavy-lift rocket. (Left) In the Vehicle Assembly Building at Kennedy Space Center, cranes lift the cage containing an Approach and Landing Test Assembly pod off the rear of *Endeavour*. (Above) This artist's rendition of the space shuttle is on exhibit at the Kennedy Space Center Visitor Complex.

# Space Shuttle
# Missions List

| April 12, 1981 | STS-1 | Columbia | 02d 06h | First shuttle flight |
| Nov. 12, 1981 | STS-2 | Columbia | 02d 06h | First flight of a reusable space vehicle |
| Mar. 22, 1982 | STS-3 | Columbia | 08d 00h | Only landing at White Sands MR |
| Jun. 27, 1982 | STS-4 | Columbia | 07d 01h | First Department of Defense payload |
| Nov. 11, 1982 | STS-5 | Columbia | 05d 02h | Multiple comsat deployments |
| April 4, 1983 | STS-6 | Challenger | 05d 00h | First space shuttle EVA |
| June 18, 1983 | STS-7 | Challenger | 06d 02h | First US woman in space: Sally Ride |
| Aug. 30, 1983 | STS-8 | Challenger | 06d 01h | First African-American: Guy Bluford |
| Nov. 28, 1983 | STS-9 | Columbia | 10d 07h | First Spacelab mission |
| Feb. 3, 1984 | STS-41-B | Challenger | 07d 23h | First untethered spacewalk |
| April 6, 1984 | STS-41-C | Challenger | 06d 23h | Solar Max servicing |
| Aug. 30, 1984 | STS-41-D | Discovery | 06d 00h | Test of OAST-1 Solar Array |
| Oct. 5, 1984 | STS-41-G | Challenger | 08d 05h | First female spacewalk: Kathy Sullivan |
| Nov. 8, 1984 | STS-51-A | Discovery | 07d 23h | Recovery of Palapa B2 and Westar VI |
| Jan. 24, 1985 | STS-51-C | Discovery | 03d 01h | First classified DoD mission |
| April 12, 1985 | STS-51-D | Discovery | 06d 23h | First flight of sitting politician: Jake Garn |
| April 29, 1985 | STS-51-B | Challenger | 07d 00h | Spacelab mission |
| June 17, 1985 | STS-51-G | Discovery | 07d 01h | First Arab in space: Sultan Al Saud |
| July 29, 1985 | STS-51-F | Challenger | 07d 22h | Spacelab mission |
| Aug. 27, 1985 | STS-51-I | Discovery | 07d 02h | Multiple comsat deployments |
| Oct. 3, 1985 | STS-51-J | Atlantis | 04d 01h | Second classified DoD mission |
| Oct. 30, 1985 | STS-61-A | Challenger | 07d 00h | Spacelab-D1 |
| Nov. 26, 1985 | STS-61-B | Atlantis | 06d 21h | Multiple comsat deployment |
| Jan. 12, 1986 | STS-61-C | Columbia | 06d 02h | Flight of US Representative Bill Nelson |
| Jan. 28, 1986 | STS-51-L | Challenger | 01m 13s | Loss of vehicle and crew |
| Sept. 29, 1988 | STS-26 | Discovery | 04d 01h | First post-Challenger flight |
| Dec. 2, 1988 | STS-27 | Atlantis | 04d 09h | Third classified DoD mission |
| March 13, 1989 | STS-29 | Discovery | 04d 23h | Tracking and Data Relay Satellite |
| May 4, 1989 | STS-30 | Atlantis | 04d 00h | Magellan Venus probe deployment |
| Aug. 8, 1989 | STS-28 | Columbia | 05d 01h | Fourth classified DoD mission |
| Oct. 18, 1989 | STS-34 | Atlantis | 04d 23h | Galileo Jupiter probe deployment |
| Nov. 22, 1989 | STS-33 | Discovery | 05d 00h | Fifth classified DoD mission |
| Jan. 9, 1990 | STS-32 | Columbia | 10d 21h | SYNCOM IV-F5 satellite deployment |
| Feb. 28, 1990 | STS-36 | Atlantis | 04d 10h | Sixth classified DoD mission |
| April 24, 1990 | STS-31 | Discovery | 05d 01h | Hubble Space Telescope deployment |
| Oct. 6, 1990 | STS-41 | Discovery | 04d 02h | Ulysses solar probe deployment |
| Nov. 15, 1990 | STS-38 | Atlantis | 04d 21h | Seventh classified DoD mission |
| Dec. 2, 1990 | STS-35 | Columbia | 08d 23h | ASTRO-1 observatory |
| April 5, 1991 | STS-37 | Atlantis | 05d 23h | Compton Gamma Ray Observatory |
| April 28, 1991 | STS-39 | Discovery | 08d 07h | First unclassified DoD mission |
| June 5, 1991 | STS-40 | Columbia | 09d 02h | Spacelab mission |
| Aug. 2, 1991 | STS-43 | Atlantis | 08d 21h | Tracking and Data Relay Satellite |
| Sep. 12, 1991 | STS-48 | Discovery | 05d 08h | Upper Atmosphere Research Satellite |
| Nov. 24, 1991 | STS-44 | Atlantis | 06d 22h | Defense Support Program satellite |
| Jan. 22, 1992 | STS-42 | Discovery | 08d 01h | Spacelab mission |

| Date | Mission | Orbiter | Duration | Notes |
|------|---------|---------|----------|-------|
| Mar. 24, 1992 | STS-45 | Atlantis | 08d 22h | ATLAS-1 science platform |
| May 7, 1992 | STS-49 | Endeavour | 08d 21h | Intelsat VI repair |
| June 25, 1992 | STS-50 | Columbia | 13d 19h | Spacelab mission |
| July 31, 1992 | STS-46 | Atlantis | 07d 23h | EURECA (European Retrievable Carrier) |
| Sep. 12, 1992 | STS-47 | Endeavour | 07d 22h | Spacelab-J, Japan-funded mission |
| Oct. 22, 1992 | STS-52 | Columbia | 09d 20h | Laser Geodynamics Satellite II |
| Dec. 2, 1992 | STS-53 | Discovery | 07d 07h | Partially classified 10th DoD mission |
| Jan. 13, 1993 | STS-54 | Endeavour | 05d 23h | Tracking and Data Relay Satellite |
| April 8, 1993 | STS-56 | Discovery | 09d 06h | ATLAS-2 science platform |
| April 26, 1993 | STS-55 | Columbia | 09d 23h | Spacelab-D2, Germany-funded mission |
| June 21, 1993 | STS-57 | Endeavour | 09d 23h | SPACEHAB, EURECA |
| Sep. 12, 1993 | STS-51 | Discovery | 09d 20h | ACTS satellite deployed |
| Oct. 18, 1993 | STS-58 | Columbia | 14d 00h | Spacelab mission |
| Dec. 2, 1993 | STS-61 | Endeavour | 10d 19h | Hubble Space Telescope servicing |
| Feb. 3, 1994 | STS-60 | Discovery | 07d 06h | SPACEHAB, Wake Shield Facility |
| Mar. 4, 1994 | STS-62 | Columbia | 13d 23h | Microgravity experiments |
| April 9, 1994 | STS-59 | Endeavour | 11d 05h | Shuttle Radar Laboratory-1 |
| July 8, 1994 | STS-65 | Columbia | 14d 17h | Spacelab mission |
| Sep. 9, 1994 | STS-64 | Discovery | 10d 22h | Multiple science experiments; SPARTAN |
| Sep. 30, 1994 | STS-68 | Endeavour | 11d 05h | Space Radar Laboratory-2 |
| Nov. 3, 1994 | STS-66 | Atlantis | 10d 22h | ATLAS-3 science platform |
| Feb. 3, 1995 | STS-63 | Discovery | 08d 06h | Mir rendezvous, SPACEHAB |
| Mar. 2, 1995 | STS-67 | Endeavour | 16d 15h | ASTRO-2 observatory |
| June 27, 1995 | STS-71 | Atlantis | 09d 19h | First Shuttle-Mir docking |
| July 13, 1995 | STS-70 | Discovery | 08d 22h | Tracking and Data Relay Satellite |
| Sep. 7, 1995 | STS-69 | Endeavour | 10d 20h | Wake Shield Facility, SPARTAN |
| Oct. 20, 1995 | STS-73 | Columbia | 15d 21h | Spacelab mission |
| Nov. 12, 1995 | STS-74 | Atlantis | 08d 04h | 2nd Shuttle-Mir docking |
| Jan. 11, 1996 | STS-72 | Endeavour | 08d 22h | Retrieved Japan's Space Flyer Unit |
| Feb. 22, 1996 | STS-75 | Columbia | 15d 17h | Tethered satellite reflight |
| Mar. 22, 1996 | STS-76 | Atlantis | 09d 05h | Shuttle-Mir docking |
| May 19, 1996 | STS-77 | Endeavour | 10d 00h | SPACEHAB; SPARTAN |
| June 20, 1996 | STS-78 | Columbia | 16d 21h | Spacelab mission |
| Sep. 16, 1996 | STS-79 | Atlantis | 10d 03h | Shuttle-Mir docking |
| Nov. 19, 1996 | STS-80 | Columbia | 17d 15h | Wake Shield Facility; ASTRO-SPAS |
| Jan. 12, 1997 | STS-81 | Atlantis | 10d 04h | Shuttle-Mir docking |
| Feb. 11, 1997 | STS-82 | Discovery | 09d 23h | Hubble Space Telescope servicing |
| April 4, 1997 | STS-83 | Columbia | 03d 23h | Truncated due to fuel cell problem |
| May 15, 1997 | STS-84 | Atlantis | 09d 05h | Shuttle-Mir docking |
| July 1, 1997 | STS-94 | Columbia | 15d 16h | Spacelab mission |
| Aug. 7, 1997 | STS-85 | Discovery | 11d 20h | CRISTA-SPAS |
| Sep. 25, 1997 | STS-86 | Atlantis | 10d 19h | Shuttle-Mir docking |
| Nov. 19, 1997 | STS-87 | Columbia | 15d 16h | Microgravity experiments, SPARTAN |
| Jan. 22, 1998 | STS-89 | Endeavour | 08d 19h | Shuttle-Mir docking |
| April 17, 1998 | STS-90 | Columbia | 15d 21h | Spacelab mission |
| June 2, 1998 | STS-91 | Discovery | 09d 19h | Last Shuttle-Mir docking |
| Oct. 29, 1998 | STS-95 | Discovery | 08d 21h | SPACEHAB; John Glenn flies again |
| Dec. 4, 1998 | STS-88 | Endeavour | 11d 19h | First Shuttle ISS assembly flight |
| May 27, 1999 | STS-96 | Discovery | 09d 19h | ISS supply flight |

(Continued on Page 156)

| Date | Mission | Orbiter | Duration | Notes |
|---|---|---|---|---|
| July 23, 1999 | STS-93 | Columbia | 04d 22h | Chandra X-ray Observatory deployed |
| Dec. 19, 1999 | STS-103 | Discovery | 07d 23h | Hubble Space Telescope servicing |
| Feb. 11, 2000 | STS-99 | Endeavour | 11d 05h | Shuttle Radar Topography Mission |
| May 19, 2000 | STS-101 | Atlantis | 09d 21h | ISS supply flight |
| Sep. 8, 2000 | STS-106 | Atlantis | 11d 19h | ISS supply flight |
| Oct. 11, 2000 | STS-92 | Discovery | 12d 21h | ISS assembly flight |
| Nov. 30, 2000 | STS-97 | Endeavour | 10d 19h | ISS assembly flight |
| Feb. 7, 2001 | STS-98 | Atlantis | 12d 21h | ISS assembly flight |
| Mar. 8, 2001 | STS-102 | Discovery | 12d 19h | ISS supply, crew rotation |
| April 19, 2001 | STS-100 | Endeavour | 11d 21h | ISS assembly flight |
| July 12, 2001 | STS-104 | Atlantis | 12d 18h | ISS assembly flight |
| Aug. 10, 2001 | STS-105 | Discovery | 11d 21h | ISS supply, crew rotation |
| Dec. 5, 2001 | STS-108 | Endeavour | 11d 19h | ISS supply, crew rotation |
| Mar. 1, 2002 | STS-109 | Columbia | 10d 22h | Hubble Space Telescope servicing |
| April 8, 2002 | STS-110 | Atlantis | 10d 19h | ISS assembly flight |
| June 5, 2002 | STS-111 | Endeavour | 13d 20h | ISS supply, crew rotation |
| Oct. 7, 2002 | STS-112 | Atlantis | 10d 19h | ISS assembly flight |
| Nov. 23, 2002 | STS-113 | Endeavour | 13d 18h | ISS assembly flight |
| Jan. 16, 2003 | STS-107 | Columbia | 15d 22h | SPACEHAB; Loss of vehicle and crew |
| July 26, 2005 | STS-114 | Discovery | 13d 21h | Flight safety evaluation/ISS supply |
| July 4, 2006 | STS-121 | Discovery | 12d 18h | ISS Flight ULF1.1: Supply, crew rotation |
| Sep. 9, 2006 | STS-115 | Atlantis | 11d 19h | ISS assembly flight |
| Dec. 9, 2006 | STS-116 | Discovery | 12d 21h | ISS assembly flight, crew rotation |
| June 8, 2007 | STS-117 | Atlantis | 13d 20h | ISS assembly flight, crew rotation |
| Aug. 8, 2007 | STS-118 | Endeavour | 12d 18h | ISS assembly flight |
| Oct. 23, 2007 | STS-120 | Discovery | 15d 02h | ISS assembly flight, crew rotation |
| Feb. 7, 2008 | STS-122 | Atlantis | 12d 18h | ISS assembly flight, crew rotation |
| Mar. 11, 2008 | STS-123 | Endeavour | 15d 18h | ISS assembly flight, crew rotation |
| May 31, 2008 | STS-124 | Discovery | 13d 18h | ISS assembly flight |
| Nov. 14, 2008 | STS-126 | Endeavour | 15d 20h | ISS assembly flight, crew rotation |
| Mar. 15, 2009 | STS-119 | Discovery | 12d 19h | ISS assembly flight |
| May 11, 2009 | STS-125 | Atlantis | 12d 21h | Last Hubble Space Telescope servicing |
| July 15, 2009 | STS-127 | Endeavour | 15d 16h | ISS assembly flight |
| Aug. 28, 2009 | STS-128 | Discovery | 13d 21h | ISS assembly flight |
| Nov. 16, 2009 | STS-129 | Atlantis | 10d 19h | ISS assembly flight |
| Feb. 8, 2010 | STS-130 | Endeavour | 13d 18h | ISS assembly flight |
| April 5, 2010 | STS-131 | Discovery | 15d 03h | ISS assembly flight |
| May 14, 2010 | STS-132 | Atlantis | 11d 18h | ISS assembly flight |
| Feb.24, 2011 | STS-133 | Discovery | 12d 19h | ISS assembly flight |
| May 16, 2011 | STS-134 | Endeavour | 15d 18h | ISS assembly flight |
| July 8, 2011 | STS-135 | Atlantis | 12d 18h | Logistics Module Raffaello |

# Photo Credits: